paving the way

how we achieve clean, safe and attractive streets

a research project commissioned by CABE and ODPM

produced by Alan Baxter & Associates in association with EDAW

Thomas Telford

cabe

OFFICE OF THE
DEPUTY PRIME MINISTER

CABE, the Commission for Architecture and the Built Environment, is the nation's
champion for better places, places which
- work better
- feel better
- are better

We believe that decent homes, work places, shops, schools and hospitals are
everyone's basic right, a right worth fighting for. CABE uses its skills and resources to
campaign for a better quality of life for people and communities across England. We do
this through a rich mix of campaigning, researching, instigating and assisting with
technical expertise and opinion.

Commission for Architecture and the Built Environment (CABE)
Tower Building
11 York Street
London SE1 7NX
Telephone 020 7960 2400
www.cabe.org.uk

Following the reorganisation of the government in May 2002, the responsibilities of
the former Department for Transport, Local Government and the Regions (DTLR) in
this area were transferred to the Office of the Deputy Prime Minister

Office of the Deputy Prime Minister (ODPM)
Eland House
Bressenden Place
London SW1E 5DU

Tel: 020 7944 3000
www.odpm.gov.uk

Throughout this report, reference is made to DTLR as the relevant Government
department at the time of the research

Copies of this report are available on CABE's website at www.cabe.org.uk
Further copies of this report are available from:

Thomas Telford

The Customer Services Department
Thomas Telford Limited, Units I/K
Paddock Wood Distribution Centre
Paddock Wood, Tonbridge
Kent TN12 6UU
Tel: 020 7665 2464
Fax: 020 7665 2245

www.thomastelford.com

ISBN 07277 3140 8

Design by Kneath Associates
Printed in Great Britain by Latimer Trend

Foreword

For far too long the street has been a neglected part of our everyday environment.

Streets are not just conduits for transport. Streets are where people meet each other. Streets define neighbourhoods. Streets have individual characters and qualities.

So why do all the rules and regulations, all the codes and manuals, still treat the street as if it were the exclusive province of the motorist?

Many people feel that they have lost ownership of their own streets. Public frustration with this imbalance has led to a growing and vocal concern. CABE's own research has shown that local residents care passionately about the quality of their street environment, but feel helpless to do anything about declining standards. Local communities want to be more involved in their own street schemes – to reap the social benefits, not least in terms of greater feelings of personal and community safety.

In *Paving the way – how we achieve clean, safe and attractive streets,* CABE sets out a clear agenda to create better designed, better managed streets, which satisfy the needs of all who use them. By close scrutiny of a set of case studies, we are able to issue a number of challenges to the institutions who really control our streets.

The Government will have to remove the inherent bias towards car users' interests in national design and regulatory guidance. We set out how they can do this, step-by-step.

The Government must also get a grip on the many vested interests that screw up our streetscapes, including the utility companies and the advertising industry.

Local authorities urgently need to co-ordinate and integrate their own management responsibilities for streets. In some areas it works well. In other areas the arrangements are a shambles. We need to bring the rest up to the standard of the best.

All professionals dealing with streets should have higher quality design training to better equip them for the task. For too long, planners and engineers have got away with second-rate interventions, using low grade materials and poor workmanship.

CABE believes that overcoming all these barriers is essential to hitting higher environmental standards.

Change is achievable. Creating better street environments is, first and foremost, about changing attitudes and ways of working, so that people are put first. CABE is committed to working with partners across the public and private sectors to ensure the goals set out in this report are achieved.

Sir Stuart Lipton
Chairman
CABE

Sir Stuart Lipton
Chairman
CABE

Contents

Management processes

There is a confusing division of responsibilities, both between different local authorities and also among different departments within authorities. These divisions result in the lack of a coherent design and management ethos for looking after streets. In urban areas, the relevant local (or unitary) authority should have clear responsibility for all roads and streets. Within local authorities, a cross-sectoral, integrated approach to the management of streets should be widely used.

Design detail and training

Specialist urban street design still receives insufficient representation in both training and local policy-making. Special elements on streetscape design should be incorporated into the professional training courses of all those who ultimately have responsibility for streets. Local transport plans and development plans (or their policy successors) should contain specific, integrated strategies for improving and maintaining the streetscape.

Control of utility companies

One of the most commonly-voiced complaints about street quality concerns the deleterious effect of utility companies' street works whereby new schemes are often undermined by subsequent poor quality interventions. The introduction of a charging regime for utilities street works, and a strengthening of local authorities' powers to fine utilities for inadequate highway reinstatements should reduce this disruption.

Long-term care

Ultimately the long-term well being of our streets equally depends on the degree of public involvement in their management and care. Local authorities should encourage and promote community commitment to street improvement projects, such as the Safe Routes to Schools Programme, and through advice and seed-funding support the creation of local trusts for that purpose.

Finally, the report raises the central issue of how we behave towards each other as shared users of the street. There is considerable confusion about the relative rights of pedestrians, cyclists and motorists, usually with the result that vehicles dominate the street so pedestrians (especially the disabled) feel constantly vulnerable. The Highway Code should be rewritten, placing greater emphasis on the multiple functions of streets, thereby championing streets for the benefit of all users not just motorists.

This report shows that generally the right design policies for streets already exist: the problem is how to implement them. The recommendations could all be introduced within a short timescale of about three years. What is required is a change of attitudes and priorities amongst politicians, policy-makers, practitioners and also the public. Importantly, the prospect of stopping the ongoing neglect of our streets is achievable. The effort and resources required are modest compared with other social priorities, yet the wider benefits to community life of creating better streets are far-reaching.

Introduction

1.0

1.0 Introduction

How to improve the quality of our streets is a key issue for everyone who cares about our towns and cities. Through neglect or oversight, or because the wrong policies have been applied, the general standard of our streets has deteriorated. We expect our local authorities to care for them, but it is not clear what their priorities should be, or how their responsibilities should be coordinated and funded.

The key issue is not so much the absence of design advice, but the quality of that advice and the way it has been used. There exists a plethora of documents relating to the street, from national statute law, regulations and policy guidance to local design guides but, despite the increasing number of documents, it seems that the general condition of our streets has remained the same or has even worsened.

This study was commissioned in May 2001 by CABE with the support of the DTLR. Its main aim is to analyse why there is a failure to create and maintain quality streetscapes. In other words, this is not yet another design guide, though it is founded on certain assumptions about the design of streets. Instead, its main focus is the process of decision making that has produced the kinds of street we see in towns and cities up and down the country. It endeavours to get behind the finished product to understand how and why it was created, and how it is cared for — and by doing so, to identify what the impediments are to the creation of good streetscapes.

So this study is distinguished from previous reports about streets and street design because it deals primarily with the management of the design process. In addition, it is unusual because it is based on detailed empirical evidence taken from twelve case studies. Each sample street has been closely analysed, using a consistent methodology, and the relevant local authorities and other bodies have been questioned about their work to the street. The findings and recommendations of this report are validated by reference to these studies.

This report has been prepared under the guidance of a steering group appointed by CABE, whose members are listed in Appendix 6. In addition, there have been detailed consultations with a large number of key organisations and individuals, who have been particularly helpful in the selection of case studies and in reviewing the issues and recommendations arising from those studies. In some areas the consultees have drawn attention to matters which were not apparent from the case studies, but which merit inclusion in the overall investigation. The names of the consultees are also listed in Appendix 6.

Chapter 2 looks at why streets are important to everyone's life and well-being, and how they are managed. Chapters 3 and 4 deal with the case studies, first by describing the methodology adopted in their selection and analysis, and then by reviewing the main issues which they raise.

The recommendations presented in Chapter 5 cover many aspects of the existing procedures and advice by which streets are designed and managed. They are intentionally confined to recommendations that can be brought forward and implemented within a realistic time period.

Streets are both the most central and the most multifarious aspect of the public realm, and it is the fragmentation of their treatment which is a major problem. The aim of this study is to show their treatment can be radically improved.

The quality of our streets

2.0

2.0 The quality of our streets

Streets are the most common feature of our towns and cities: they are the veins which allow places of every shape and size to function. They exist not just for movement, but as a space that everyone shares. So pervasive are they that they constantly run the risk of being taken for granted, treated as something that can always be relied upon however much they are abused or neglected. It is assumed that they can be depended on at all times and in all conditions.

The starting point for this study is the realisation that streets cannot take care of themselves. Because of their many functions they have always been the subject of potential conflicts, and this is truer than ever in our unprecedently complex society. It is increasingly difficult to know who is responsible for their different aspects, and how they relate to each other.

This chapter sets the scene for the analysis that follows. It deals with the role of streets, and why they matter in how we live and interact. It then discusses what we mean by quality in streetscape. Finally, it outlines how the management of streets has evolved, to create the complex system for their design and care which exists today.

2.1 What streets are for

Whereas words such as road (from the Anglo-Saxon ride) suggest movement from one place to another, the word street (from the Latin sternere, meaning to pave) suggests an area for public use but not exclusively devoted to circulation. The street is, by definition, a multi-functional space, providing enclosure and activity as well as movement. Its main functions are:

- circulation, for vehicles and pedestrians

- access to buildings, and the provision of light and ventilation for buildings

- a route for utilities

- storage space, especially for vehicles

- public space for human interaction and sociability; everything from parades and protests to chance encounters

Virtually all streets in urban areas perform all of these functions, and often the balance between them will vary along the length of the street.

Ideally, all these facets of the street can successfully coexist, but all too often it is one function (especially the movement of vehicles) which has been allowed to dominate.

Getting the balance right at the right place is critical because streets are the most important part of the public realm, and thus are fundamental to how we live together in towns and cities. They influence our lives at the functional level (how we get around) but also in how we relate to each other and to public authorities. They are the testing-ground for how we, as individuals, share the citizenship of the places where we live and work. A famous court case on the use of the highway has summarised the duties which arise from the sharing of this common space:

> *'The law relating to the user of highways is in truth of the law of give and take. Those who use them must in doing so have reasonable regard to the convenience and comfort of others, and must not themselves expect a degree of convenience and comfort only obtainable by disregarding that of other people. It is the price they pay for the privilege of obstructing others.'*
>
> (*Harper v. Haden & Sons Ltd 1932*)

2.2 Defining quality streetscape

The term 'streetscape' refers to the design quality of the street and its visual effect, particularly how the paved area (carriageway and footway) is laid out and treated. Obviously the buildings and other features which enclose the street are crucial to its character, but generally they are in private hands and are dealt with through different categories of statute law and guidance. What matters in this study are the parts of the street that are in shared public use.

Most people respond instinctively to the quality of streets, for instance in deciding where to live and spend their time. Part of their response relates to building uses and activities, but the overall feel of the street matters just as much. Among those who have any kind of responsibility for the street, such as the police, the local authority and property agents, it is common knowledge that some streets are successful and others are not.

Some people will put up with streets of any type or condition, but given a choice most people know what they like — their preference helps shape the character of different towns and cities. If looked at in detail, their common experience concurs with professional opinion about what constitutes a successful street.

For the purposes of this study, certain broad assumptions have been made about streets and streetscape. The main attributes which are listed here are based on the growing literature about streets and streetscape, and on our discussions with the consultees for the project.

The main indicators of quality, which are the test of successful streetscape, can be listed under six headings:

- Comfortable and safe for pedestrians and the disabled

- A street designed to accommodate all sorts of functions, not dominated by any one function

- Visually simple, and free of clutter. Regardless of whether a street is a straightforward or complex space, what matters is the simplicity and clarity of its paving, street furniture, lighting and landscaping

- Well cared for, and where utilities or 'extraneous' advertising are subordinate to all other street functions

- Sympathetic to local character and activity context, in design and detail

- Making appropriate ordered provision for access, deliveries and storage of vehicles

These are the attributes which have been used in analysing the various streets covered in the case studies.

2.3 How streets are designed and cared for

Historically, streets were the responsibility of the building owners along the street or the local parish. In some places separate Paving Commissions were set up to deal uniformly with streets in a particular area. For new developments, such as the Bloomsbury district of London, the consistent quality of the streets was due to the control exercised by the ground landlord in the design and layout, and, subsequently, the supervision of Paving Commissioners.

Ultimately this rather ad hoc system of local control was clarified, notably in the 1888 Local Government Act, which designated newly-created local authorities as highway authorities. The 1888 Act was also significant because it heralded central government involvement in the design and care of streets through central grant-funding. Direct government responsibility for roads became a reality between the wars, when the Trunk Road Act 1936 handed strategic traffic routes to the Ministry of Transport. Ever since then responsibility for roads and streets has been shared between central and local government.

Today's system of administration of streets is generally divided between national and local government under various statutes. The management responsibilities are divided for:

- motorways and trunk roads in England

- other roads in single tier authorities in England

- other roads in two tier authorities in England

- other roads within London authorities

Administration is based on statutory 'duties', i.e. absolute responsibilities, and 'powers', i.e. selectable responsibilities derived from a mix of highway, traffic planning, environmental protection and local government legislation. Details of these powers and responsibilities are shown in Appendix 2.

In the construction of new roads or streets, the responsibility for the completed works passes from the developer to the local authority under the adoption procedures laid down in the Highways Act 1980, section 38. These procedures have a major influence on the design and appearance of new street layouts (see the discussion of the Coldharbour Way case study in Section 4.1). But equally, streets that remain under the management of development companies tend to adhere to the same design principles.

The growth of regulations and guidance affecting the design of streets is a vast subject which only needs to be summarised here. From the seventeenth century, if not earlier, Acts or Bylaws were introduced aimed at making streets safer, more orderly and more visually appealing. At first the principal concern was the risk of fire, leading to the regulations in the 1667 London Building Act. Subsequently, anxieties for public health and sanitation produced prescribed street widths designed to improve daylighting and ventilation.

With the growth of motor traffic in the 1920s, attention shifted to the free-flowing movement of vehicles as the main consideration in the design of streets. The Road Improvement Act of 1925 allowed local authorities to prescribe building lines according to traffic needs, and to chamfer street corners to improve visibility at junctions. That legislation, followed by the Highways Acts of 1959 and 1980, placed increased powers in the hands of the Minister of Transport and of the highway authorities. All government advice on the design of roads and streets, notably *Design Bulletin 32* (DoT/DoE 1992) for residential roads and the *Design Manual for Roads and Bridges* (DETR 1994), reinforced the dominance of vehicle movement as the prime function of streets to the disadvantage of all other uses.

It was not until the DTLR's publication of *Places, Streets and Movement* (DETR 1998) that an attempt was made to readjust the balance in official guidance towards other considerations in street design.

The other main function of the street, which is subject to historic administrative arrangements, is its use, above and below ground, as a conduit for utilities — drainage, water, gas, electricity and telecommunications. From the medieval period onwards, but notably from the introduction of gas lighting in the 1840s, statutory undertakings have acquired the right to lay their services in the adopted highway, and to dig up the street without the need for planning or any other permission. This privilege also extends to above ground installations such as telephone lines and pillar boxes. With the deregulation of telecommunications in 1987, and the proliferation of cable and IT companies, this places extreme demands on many urban streets. The street has become the workhorse on which the servicing of towns and cities relies.

As Figure 2.1 shows, the number of controls affecting the design and use of the street (administered by different and often poorly-linked organisations) has grown significantly since 1950. The stage has now been reached when the system of regulation is beyond most people's grasp. Because of its complexity, inevitably there are potential conflicts in the system, but equally there are areas for which little

Figure 2.1: Increase in
statutory controls
affecting the street

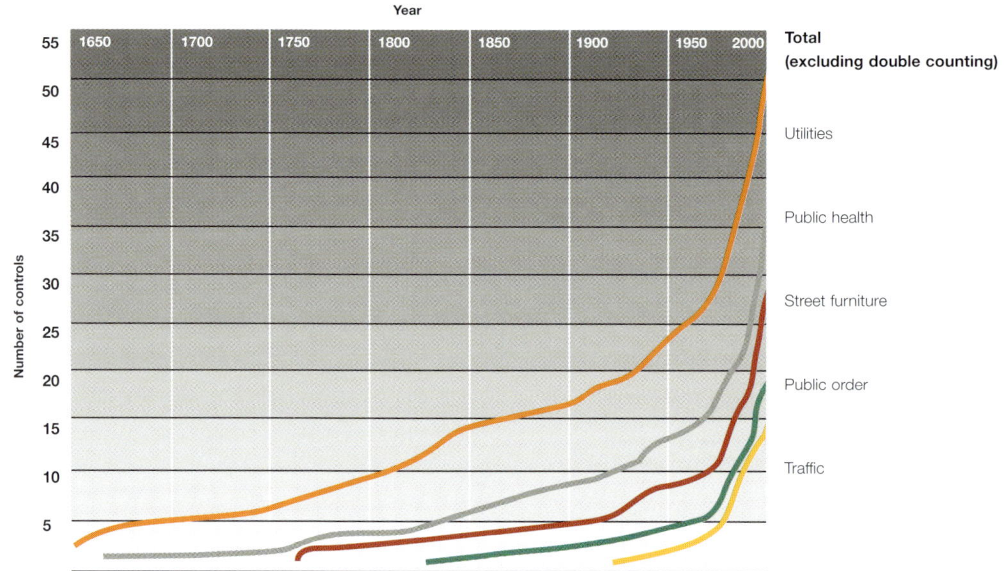

specific legislation or guidance has been provided. For instance, *Design Bulletin 32* (DoT/DoE 1992) deals specifically with residential roads and footpaths. Because other types of road, for instance urban local roads, have little specific design guidance, there is a tendency to apply inappropriate published trunk road standards to them. The controls available sometimes conflict in aspiration, and the way they are used is a question of balancing those aspirations.

The main apparatus of control and guidance, as it exists today, can be summarised under three headings as follows (for a more comprehensive list see Appendix 4).

Statutes

- Environment Protection Act 1990: collection of litter and refuse

- Highways Act 1980: powers to create or improve, and duty to maintain, the highway

- Local Government Act 2000: duty of local authorities to produce community strategies, including for the improvement of environmental well-being

- New Roads and Streetworks Act 1991: powers of coordination and inspection of utilities works. (The regulations and codes of practice under the Act are currently being reviewed)

- Road Traffic Regulation Act 1984: powers to restrict or exclude certain types of traffic

- Town and Country Planning Act 1990: definition of what constitutes development, Permitted Development Rights, advertising controls, duty of local authorities to draw up Development Plans

- Transport Act 2000: duty of local authorities to draw up Local Transport Plans

Regulations

- Control of Advertisement Regulations 1992, amended 1999

- Highways (Traffic Calming) Regulations 1993

- Regulations under the new Roads and Streetworks Act 1991

- Traffic Signs Regulations and General Directions 1994

Non-statutory design guidance

- *Circular Roads* (1970–2000)

- *Design Bulletin 32: Residential Roads and Footpaths* (1977; revised 1992)

- *Design Manual for Roads and Bridges* (1994)

- *Local Transport Notes* (1986–1998)

- *Places, Streets and Movement* (1998)

- Planning Policy Guidance, especially *PPG1 General Policies and Principles*; *PPG8 Telecommunications*; *PPG13 Transport*; *PPG19 Outdoor Advertising Control*

- *Secured by Design* (1994)

- *Traffic Advisory Leaflets* (1989 onwards)

- *Traffic Signs Manual* (1994)

In addition, there are British and European Standards for many types of street furniture, signs, lighting and road markings, and many local authorities have published design guides that include detailed recommendations on street design.

In conclusion, there is one particular aspect of the regulatory regime which deserves comment. In every sphere of life, the language used to describe something influences how it is perceived. Thus, the words used to describe roads and streets create a frame of mind about how they should be regarded. For the purposes of highway administration, it has been customary for the last 30 years or more to identify roads and streets through a hierarchical classification — arterial roads, district distributor, local distributor, access roads and shared surface roads. This classification system is intended to help prioritise capital and maintenance spending, and to guide development control and parking strategies. Some form of descriptive hierarchy is, of course, a necessity, but the terms used carry the presumption that the prime use of roads and streets is vehicle movement, and that presumption has had a fundamental influence on how they are designed and cared for. A different language, based on their overall character and functions rather than their vehicle capacity, could help point the way to a different kind of management regime.

Photo courtesy Gillespies

Case studies

3.0

3.0 Case studies

The analysis and recommendations in this report are founded on the empirical evidence derived from twelve case studies. Everyone has an opinion about how our streets might be improved but it is unusual for such discussion to be validated, as it is here, by detailed analysis of street design and management.

The case studies presented here have one thing in common, that the streets concerned have all been improved or altered in recent years. The examples range from bus priority improvements to the construction of a new inner urban road. Between them they represent the full spectrum of current practice in street design. And because they are all relatively new projects, it was assumed that it would be possible to trace the design and decision-making process that produced them.

Thus, the aim has been to evaluate each of the case study streets, using the tests of quality described in Section 2.2, but also to find out who was responsible for the works, what their aspirations were, and how each project was managed.

The selection of case studies was based on the consultants' knowledge, and the advice of the Steering Group and consultees. Seventy potential examples were considered, from which ten were chosen. The main criteria for selection were:

- streets where works had recently taken place

- a full range of street types, in different kinds of environment

- examples from different parts of the country, to reflect local influences and distinctiveness

Although some may be called roads, they are all streets in the sense of being of mixed function and in fairly busy areas.

During the course of the study, and after the initial ten examples had been chosen, it was decided to add two examples of streets in Historic Core Zones, one in Shrewsbury and one in Bury St Edmunds. The benefit of including these is that they have been the subject of particularly strict requirements, especially regarding local context. Thus, an added degree of care has gone into their design. But they should not be thought of as special cases, and part of the analysis was to identify how lessons from them could be applied in less self-consciously historic environments.

There is, of course, an almost infinite variety of streets around the country, and it might be thought that twelve streets is too small a sample on which to base any valid conclusions. However, because the cases that have been chosen are representative of a wide range of types and circumstance there is no reason to think that by expanding the sample fresh lessons would be learned.

Table 3.1: Case studies

Street type	Street name	Location
Inner urban local street	Cowcross Street	London
Inner urban main street	Castlegate	Nottingham
Inner urban major distributor	St James' Boulevard	Newcastle-upon-Tyne
Historic core zone	High Street	Shrewsbury
Historic core zone	Hatter Street	Bury St Edmunds
Suburban main street	Sutton High Street	Surrey
Suburban main street	Ladypool Road	Birmingham
Suburban distributor	Wellington Road North	Stockport
Suburban distributor	Belle Isle Road	Leeds
New suburban distributor	Coldharbour Way	Aylesbury
Rural town main street	Barras Street	Liskeard
Trunk road	West Street	Dunstable

A consistent methodology was applied in the study of the twelve examples, consisting of:

- an initial desktop analysis

- a site visit and visual appraisal

- a questionnaire to the relevant local authority concerning details of the street improvement project, including the management structure, design process, funding, and post-completion issues

- telephone interviews with the relevant officers

- the completion of a case study record, highlighting issues of importance to the overall study

    Each street has been appraised under the following six headings.

1	The level of comfort and safety of provision for disabled
2	The degree to which all functions are accommodated
3	Visual simplicity
4	The utilities function subordinate to all other street functions
5	Improvements fitting to character and activity of the street
6	Ordered for access and storage

A simple assessment system was adopted to gauge the relative success of the twelve examples.

A summary of the case studies is given in Appendix 1. Case study sheets give details about the street and show the results of the assessment. In addition, a limited amount of information has been gathered on streetscape management in three European cities — Marseilles, Copenhagen and Freiburg im Breisgau — and this is summarised in Appendix 3.

The knowledge acquired from the studies underpins the discussion of significant issues, which follows in Chapter 4. However, one point should be emphasised here since it relates directly to the information-gathering exercise that has been carried out for each street.

Although all the projects are relatively new, many local authorities have had difficulty in summarising how they had been designed and executed. Either staff had changed, or reorganisation had taken place, making it hard to trace how decisions had been made and by whom. Paperwork would not maintain an easily audited strategic picture of a project, but would be reliant on individuals who directed the scheme. The exception was the St James' Boulevard, Newcastle-upon-Tyne, case study where the need to argue a public inquiry required detailed retention of an overall view by the city authority. This absence of collective memory inevitably leads to a loss of collective responsibility — when no one knows why something was done, and no one reviews whether it was the right decision, or how to address it again in the future.

The issues and recommendations presented in this report relate specifically to how streets are designed and cared for. However, as our enquiries have shown, at a broader level this study raises the general issue of continuity and responsibility in local authority management.

The impediments to quality

4.0

4.0 The impediments to quality

The aim of the case studies has been to analyse the process for the design and management of streets, and to discover how that process influences the production of quality streetscape. This chapter discusses the main issues that have emerged from the case studies, focusing on the evidence about how decisions on streets are made and implemented.

The overall tenor of these findings may appear to be unduly negative. However it should be stressed that the efforts to improve the condition and appearance of our streets in recent years have generally been beneficial, especially if compared with achievements in previous decades. There is an increasing awareness of the need to improve the pedestrian environment, and to strike a better balance between the different functions of the street, and this is reflected in the schemes discussed here as well as in countless other projects around the country.

Many of the case study projects have been carried out to try and remedy the impact of previous street layouts designed mainly with vehicle movement in mind. There is now a new order of priorities, which is clearly spelt out in government planning and transport policy guidance, and the streets which have been analysed for this report reflect the influence of this agenda at the local level.

Broadly speaking, there is no lack of ideas about how to improve our streetscapes. The question is whether those ideas have reached those who are making the key decisions and how successfully they have carried them into effect. In other words, what is preventing the implementation of the new agenda? Where it is being put into practice, is what is designed and built of the highest quality?

The issues raised by each of the twelve case studies are itemised in Appendix 1. Here they are summarised under five broad headings.

4.1 Statute and design guidance

The compromises that threaten the implementation of government policy are most evident in the case studies of newly-constructed streets, through they are also apparent in alterations to existing streets. For instance, Coldharbour Way is part of Fairford Leys on the outskirts of Aylesbury, a new mixed-use development which ultimately will include some 1,800 houses and other facilities. Its masterplan has been designed to reduce car dependency through the location of facilities close to housing and the provision of good walking and cycling routes. It incorporates streets and squares laid out to create distinctive local areas, at moderately tight densities.

But however good the original intentions, the quality of the main streets at Fairford Leys — at least in the first phase to be completed — has been distorted because of demands imposed by the local Highway Authority. Thus, a square that was designed as an urban space through which traffic could pass has been debased by the addition of footway build-outs and deflections that emphasise the dominance of the carriageway and vehicle movement. Similarly, the intention to create a main street through the arrangement of building frontages, with the pavements following the building's line, has been eroded by the build-out of footways to delineate a standard carriageway. The result is a scheme in which the priority given to the demands of vehicle movement has undermined the quality of the original design.

Two other examples illustrate how good intentions can be compromised. Both relate to major distributor roads in urban areas.

St James' Boulevard, Newcastle-upon-Tyne, is a newly-created street, planned to handle traffic skirting the city centre. Here the overall concept is superior to what might have been done ten or twenty years earlier, especially in the desire to provide for pedestrians, cyclists, and mobility impaired people, but the detailed execution has relied on standard solutions which detract from the overall achievement. In particular, the use of pedestrian crossings with a staggered arrangement, including a narrow pen on a traffic island, suggest that pedestrians' needs are still regarded as an optional extra.

At Castlegate, Nottingham, the use of the same staggered arrangement, where the route crosses Maid Marion Way, immensely detracts from an otherwise carefully thought out and hard fought scheme to improve the pedestrian environment.

The decision making in each of these projects highlights four main issues:

- Existing statute law, regulations and design guidance are often out of tune with the priority now given to streetscape and the needs of pedestrians. They are still focused on standard solutions, based on traffic volumes, vehicle flows and vehicle speeds. There is an increasing awareness of the influence of vehicle speed on road design and its subsequent influence on quality of life.

- Local Highway Authorities rely on standard practice, giving primacy to vehicle movement, because of fear of prosecution for negligence if accidents occur. Use of official guidance is regarded as the clear line of defence if faced with litigation though negligence is primarily measured against a duty of care, not against guidance.

- There is confusion about the relative status of the many statutes and documents relating to street design.

- The powers granted to a local authority in their role as Highway Authority under the Highways Act 1980, including adoption procedures, are administered in ways that are often at odds with the aims of planning and urban design.

None of these are new issues. As already emphasised (Section 2.3), the apparatus of statute law and guidance reflects the importance attached to vehicle movement over the last 50 years or more. And during the same period, adoption procedures have led developers to use road and street layouts designed primarily with the car in mind.

> *This bias in the way roads and streets are discussed in official documents is so ingrained that it is difficult to imagine a time when it did not exist.*

The bias in official guidance and regulation in itself presents a fundamental problem in dealing with streets as multi-use environments. In contrast with car users, the claims of other users, and other functions of the street, are treated as secondary. The descriptive classification relied on (distributor road, local distributor, etc.) makes no allowance for looking at the street as a total entity. To cite but two examples. The now superseded TA52/87 *Design Considerations for Pelican and Zebra Crossings* suggested that on long stretches of straight or wide road the means for pedestrians to cross should be a footbridge or subway. The alternative possibility that the road itself can be redesigned for slower traffic speeds, with a pedestrian crossing at grade level, was not proposed at that time. Even current guidance, however, in LTN 1/95 and 2/95 (DETR 1995), suggests pelican crossings should not inconvenience the driver. Traffic Advisory Leaflet 12/97 on *Chicane Schemes* (DETR 1997) describes one mechanism for traffic calming, but the solutions illustrated are almost totally unsympathetic to the existing streetscape.

A shift in official policy towards a more balanced view of the street began in the 1990s, signalled by the publication of *PPG13: Transport* (DTLR 2001a) and *Places, Streets and Movement A Companion Guide to Design Bulletin 32* (DETR 1998). The latter document spells out a clear message that, 'in the making of places it is not the road layout but the relationship of buildings to each other which should be paramount' (p. 26).

However, as the case studies in this report show, that message still has to fight to be heard against what is regarded as the prescriptive guidance of earlier official notes and advice.

Thus, the first major impediment to creating better streetscape lies in statute law and guidance, and, in particular, the rigid adherence to design guidance which promotes a one-sided, vehicular view of the street.

4.2 Street design and management

Most of the case studies deal with comparatively short streets where improvements have been carried out as a single initiative. Yet even in such straightforward circumstances many of the examples reveal a confused or overlapping system of decision making, between or within authorities. The lack of clear management shows itself in the quality of the end result.

Obvious problems arise when more than one authority is responsible for a street or, as more usually happens, a street in the care of one authority crosses one looked after by another authority. The existing system of road classification, leading to a division of responsibility for different types of road or street, results in such situations which are a natural breeding-ground for conflict.

> *There is a prevailing sense that no one is in overall charge of the appearance and function of streets.*

Two examples from the case studies illustrate this. In Nottingham, the Castlegate project was conceived of by the City Council as the Planning Authority, but Nottinghamshire County Council as the Highway Authority took a major interest in it, especially where Castlegate crosses the inner ring road Maid Marion Way. The discordant design where the two streets cross reflects the lack of coordination between the aims of the two authorities, the County focused on vehicular flows and the City on pedestrian flows.

In Sutton in South London, the High Street is a typical suburban main street looked after by the local borough, but it is bisected by the A232, which is part of the Greater London Strategic Road Network administered by Transport for London. Where one crosses the other there is a total change in design treatment, including different paving and uncoordinated street furniture. Such anomalies are avoided when all streets, except motorways, are the responsibility of a single authority. Many European cities, for instance in Germany, have a clarity of administration which is reflected in the quality of their streetscape (see Appendix 3).

Where there is one controlling authority these issues ought not to arise, yet they reappear in the form of conflicts between administrative departments, or as a result of changes in personnel during a project. Many of the case studies reflect the lack of a corporate policy framework or clear management structure for dealing with street design and maintenance.

Again, Sutton High Street provides an apt illustration of this issue. There the design of the street works was entrusted to the Highways Department, but with a landscape architect introduced to try and improve the end result. In fact, what has been produced is visually confusing in its materials and details, reflecting the failure to establish a clear design strategy at the outset. At Belle Isle Road in Leeds, street improvements, drawn up as part of a project to regenerate adjoining housing, have been compromised because of a lack of a clear strategy for public transport provision.

By contrast, St James' Boulevard in Newcastle-upon-Tyne, though spoilt by the design of its pedestrian crossings, does reflect the benefits of a multi-disciplinary team approach.

Generally, the case studies illustrate one of the points that is most often made about the quality of streets in England, above all the sense that no one is in overall charge of their appearance and function. This is usually apparent in the siting and design of street furniture and signage, where things are added over the years but apparently nothing is taken away. The projects analysed in the case studies are all the product of recent initiatives, and so do not yet show the effects of that kind of neglect. Yet, even as conceived and built, many manifest the symptoms of confused, unsystematic decision-making.

Decision making is dictated by statutory duties on local authorities which affect street design and management thinking. It also reflects the priority given to using policies and powers that are present. Local authorities, strapped for resources, resort to only providing the 'must do' services rather than the 'can do if necessary'.

4.3 Design detail

The two main issues so far discussed — the application of statute law, regulations and guidance, and the management framework — are of fundamental importance to understanding the quality of streetscape design. They represent the context within which detailed decisions are made: if that context is wrong the odds are against achieving a first rate result.

But the detailed design can also make or break a scheme. That is to say, in a well-conceived project it is still immensely important that the overall aims should infuse every aspect of its design and execution. The detail should not be an afterthought, to be attended to as circumstances allow, but an integral part of the overall design process.

A number of the case studies illustrate how attention to detail can make all the difference to streetscape. There are four ways that quality can be compromised or eroded:

- **Lack of a clear design ethos.** Street furniture and fittings and the design of paving, ideally, should be simple and consistent. At Dunstable in Bedfordshire, for example, quite the opposite has happened. Some bollards and signage are historical pastiche, while other features, including seating, are modern. The coloured patterns in the paving bear no relation to the street form and usage. The visual effect, after much effort and expenditure, is confusing and illogical

- **Conflicting design solutions.** Because of the varied functions of streets, their design has to answer many aims. Often it appears that each function has been dealt with in isolation, resulting in a lack of coordinated design. Thus, on Belle Isle Road in Leeds, the speed cushions introduced for traffic calming conflict with the routes selected for pedestrian crossings. The new on-street parking is equally arbitrary. Similarly, Sutton High Street shows the confusing overlay of rival design priorities

- **Inappropriate signage.** The need for clear signage is an obvious aspect of street design. The use of a consistent graphic language of lettering and symbols contributes to the good appearance of the street and to public safety. But signs, however well-designed, can cause immense harm to the character of a street if they are badly sited or of an inappropriate scale. They are sized on the basis of traffic speed to facilitate driver legibility. As a case in point, at Wellington Road North in Stockport a well-conceived improvement to bus, cycling and pedestrian facilities has been marred by the use of traffic signs which are of an excessive scale for a residential area. No thought appears to have been given to varying traffic management and signage to suit the local context. Other case studies contain similar examples of the use of standard-sized signs regardless of the local environment

  In Historic Core Zones, as at Bury St Edmunds, the standard signage has been imaginatively adapted to suit the local context.

In mainland Europe signage is often adapted to suit the local contest. Some places have gone as far as removing all signage. For instance, at Oosterwalde in Holland, two busy road junctions (up to 10,000 vehicles/day) have been redesigned without signs, lines or signals. Pedestrians and vehicles mix easily without accidents, establishing eye contact to negotiate right of way.

- **Neglect of the local vernacular.** Streets, as much as buildings, reflect local traditions in layout, materials and features, all of which merit respect in alterations and new design.

  Though the patterns of local distinctiveness are stronger in some parts of the country than others, wherever they are evident they should be acknowledged. Cornwall is one area that has a distinctive streetscape vernacular, and it is surprising that the improvements in Liskeard, subject of one of the case studies, fail to make more use of local characteristics. The new works here have a palette of materials that blends successfully in colour to the local architecture, and budgets have allowed for some use of granite bollards. But the traditional paving pattern has not been followed, and new materials have been crudely imposed. For instance, footway widenings have left the old kerbs in place making them appear as afterthoughts, and the new kerbs and paving flags are unsympathetic to the character of the original paving.

Regarding matters of detail, some of the case studies show just how much can be achieved. In the two Historic Core Zone projects analysed — at Bury St Edmunds and Shrewsbury — there has been an above-average attempt to create an effective layout using natural or more carefully blended materials, and to control the design and location of signage while implementing complex traffic management schemes. The same effect of clarity and consistency has been produced at Cowcross Street, London, in a less historically self-conscious manner. What these projects suggest is that the institutional barriers to good design can be surmounted when it is felt that special circumstances exist.

> *Every street is special, not just ones in historically significant areas.*

The question this raises is how to instil an attitude that every street is special, not just ones in historically significant areas, and that care of the streetscape should be given high priority. This points to the subject of knowledge and training, and the skill of those involved in street design and management. The case studies confirm the conclusions reached in other reports, such as the *Urban Design Skills Working Group Report* (CABE, 2001), in particular that streets (and the public realm generally) are given insufficient prominence in professional and vocational education; and that those who have the correct training and skills are not being given the right degree of responsibility in local government. This lack of design talent, and the failure to exploit the talent that does exist, shows itself in the condition of streets throughout the country.

4.4 Licensed operators

In almost every new street or street improvement, a recurrent issue concerns the powers granted to licensed operators to undertake works in the street. With so many demands on their resources, local Highway Authorities are often not able to monitor utilities adequately. The most frequently cited impacts are:

- works by utilities to the highway and footway, using rights granted by various statutes, especially the New Roads and Streetworks Act 1991 — this includes the siting of telephone boxes and other equipment, such as junction boxes

- advertising, bill boards and fly posting, all of which come under the Town and Country Planning (Control of Advertisements) Regulations 1992

The most conspicuous problem arises from works by the utilities, especially in streets where care has been taken to complete a coordinated, high quality scheme. The disruption caused by utilities works is a common subject of public complaint: indeed there is probably no aspect of the public realm about which people feel so strongly. But apart from disruption, the visual aftermath of such works is often a permanent reminder of the casual way utilities treat our streets. This problem has become far worse in recent years with the proliferation of rival telecommunications companies, whose operations are subject to little overall supervision and effective enforcement of regulations.

> *Local authorites appear powerless to prevent utility works spoiling the streetscape.*

What matters here is the quality of reinstatement after opening-up works in the street have been completed. All too often carefully laid finishes (granite setts or stone paving) are replaced by poorly backfilled trenches and crudely applied tarmac, leaving a scar across the original design. In one of the case studies, at Cowcross Street in London, the completion of a major improvement exercise was followed by no less than 16 separate openings made by the utilities, often poorly reinstated and, thus, utterly at odds with the aim of the street improvement. After two years, permanent reinstatement has not been carried out.

Under the provisions of the New Roads and Streetworks Act, local authorities are funded to inspect 30% of reinstatement works (currently 6% at five work stages, although it is being considered for revision). The evidence of the case studies suggests that this regime of inspection is ineffective, or that the enforcement measures currently in existence are often not fully utilised due to their inadequate funding and therefore low prioritisation. Current legislation does not allow local authorities to recover inspection costs for all reinstatement works. The City of London and Westminster collected £750,000 in fines since the 2001 fining Regulations were introduced, demonstrating that the effort spent collecting fines has its compensations.

In contrast to the English experience, the well cared for feel of streets in many European cities is partly the result of more stringent control of utility streetworks, both in the planning of works and in the supervision of reinstatement once works have been completed.

The siting of above ground equipment can be equally detrimental to the quality of a street. Among the case studies, Sutton High Street provides a typical example of the casual siting of telephone kiosks and control boxes. Similarly, at Dunstable no thought appears to have been given to the location and impact of telecommunications equipment (for which planning permission is not required). By contrast, the Shrewsbury case study includes an example of a control box carefully sited within a wall opening — another instance where a historic environment provides a lesson which could be widely applicable elsewhere.

Advertising was cited as a major issue by the consultees to this project, who drew attention, in particular, to:

- the potential of well-designed advertising to contribute to the quality of streetscape in certain locations
- the deleterious effect of advertising on street furniture — in this respect there may be a conflict between the planning aspirations of a local authority and the revenue potential from street furniture advertising
- the impact of fly-posting and shroud advertising (i.e. the unauthorised use of a building scaffold for advertising)

These issues have not come to the forefront in any of the case studies, but have been considered as part of this project because of their general relevance to streetscape quality.

4.5 Long-term care of the street

The twelve case studies all feature projects that have been completed in recent years, and thus provide less evidence on the long-term care of the street than on other issues. However, as already emphasised, the studies do include examples of the harm caused by utilities even within months of a project's completion. These are salutary reminders of the fact that planning for aftercare needs to be fully assessed in the design process, and steps for continued care instigated promptly the moment construction is finished.

The value of initiatives to create or improve quality streetscapes depends significantly on the attention given to long-term control and maintenance. Even the best-conceived scheme can deteriorate within an alarmingly short period if these aspects are neglected. The care of the street raises issues of management and detailed design, which have already been discussed. If, for instance, a highway engineer fails to appreciate what his or her colleagues were attempting to achieve in carrying out improvements, those good intentions can easily be compromised by subsequent alterations (in some respects this is what has happened at the High

Street, Shrewsbury, and Sutton High Street). Conversely, the original design should take full account of how the street will be cared for in the future, for instance in the choice of materials and street furniture.

But where long-term care is concerned there is one further consideration, which is the sense of shared responsibility for the street. The neglect of our streets arises partly from the fact people feel no commitment towards them: they are seen as a local authority problem, not theirs. The involvement of owners and users in deciding how a street should be looked after, and what improvements it requires, is potentially a way to rekindle an interest in the public realm. It is also a way to ensure that what is decided about a street is in line with people's aspirations, and is not just the application of standard ideas unrelated to the local context.

Recommendations

5.0

5.0 Recommendations

The evidence of the twelve case studies has helped highlight the issues of streetscape quality that most need to be addressed. The studies show that there is a widespread ambition to improve the quality of our streetscapes, in places of every size and character. But this ambition has been impeded by numerous barriers to quality, with the result that what has been achieved generally falls short of the ideal. As an assessment of the current standard of streetscape design and management, the studies reveal a disappointing picture. There is no reason to think that a wider selection would have yielded notably different results.

Put simply, even with the best intentions the quality of streetscapes is often compromised by the rigid application of highway engineering solutions, the lack of a coordinated approach to design, and a failure of design detail. And even where a well-conceived scheme has been successfully executed, it most probably will be debased by works undertaken by utility companies.

The recommendations that follow have been devised to deal with the underlying, generic issues which have emerged from the case studies, as well as from discussions with the project consultees. They are presented in essentially the same order as the discussion of issues in Chapter 4, so that each recommendation can be traced to the empirical data on which it is based. The only exception in this respect concerns the recommendations on the *Highway Code*, which is dealt with separately as the final recommendation.

Inevitably, some recommendations are more significant than others, and the order in which they should be addressed is also important. Some could be implemented incrementally, as part of a programme, some are direct actions. Under each heading the recommendations are presented according to the time-scale and difficulty of their implementation. In addition, at the end of this chapter there is a short discussion of the process for bringing forward the recommendations. This is not intended to specify the exact mechanisms for implementing each proposal, but rather is meant to indicate how the priorities emerging from this report relate to other initiatives and research.

5.1 Statute and design guidance

> *Recommendation 1: A succinct summary should be published, detailing the statutes, regulations and design guidance affecting streetscape design and management. This summary should distinguish clearly between advisory and mandatory documents.*

At present there is considerable confusion about the relative status of the many government documents relating to streets, in particular what is required by law and what is advisory, with the result that standard or inappropriate solutions are adopted in many instances (see Appendix 4 for a list of the relevant documents). This proposed summary (which should be regularly updated) would help clarify how engineers and urban designers should use each document, and will provide a high-level policy direction on street design and liveability. It will help draw attention to streetscape and public realm as distinct responsibilities, and prevent the use of particular guidance in isolation from others. The preparation of this summary document will constitute an 'early win', indicating that the issues raised in this report are being addressed.

> *Recommendation 2: Highway authorities should, under Best Value, establish an audit trail for design decisions affecting the streetscape, to show how design guidance, people's needs and vehicle movements have been accommodated.*

The aim of this proposed audit trail is twofold:

- To enable local authorities to demonstrate that they have acted 'reasonably', if faced by liability claims. Often the main reason cited for the use of a standard (vehicle priority) solution is that such a solution, however inappropriate to the local context, is thought to offer a line of defence in a Court of Law. As an alternative, the audit trail offers a means of demonstrating that due care was taken in the design of a streetscape project, especially when that design is tailored to particular local circumstances

- To act as a record of aims and objectives about both how and why decisions were made, and to demonstrate that all users (including the disabled) have been considered in decision making. There will be a reference point for future projects on the same street

Audits and risk assessments for safety and cost are already implemented by many authorities but they are snap shot views rather than long-term tools. There is no current requirement to retain this information or make it public other than as a standard duty of care.

> *Recommendation 3: Existing guidance on highway design should be rewritten to bring it in line with Government policy on design, sustainability and the urban realm.*

This proposal is not intended as a challenge to the detailed technical content of the highway design literature (*Design Bulletin 32, The Design Manual for Roads and Bridges, Traffic Advisory Leaflets,* and *Local Transport Notes*). Instead it is aimed at the failure of such documents to acknowledge the multiple functions of the street, and the balance to be drawn between different functions at particular locations. Generally, the highway design literature focuses mainly on the street as a route for traffic flows, and it gives insufficient or no attention to other forms of movement (especially pedestrian movement) and other activities in the street. In addition, it uses a language to describe roads and streets (district distributor, local distributor, etc.) which is at odds with the overall functions of the street. As part of the review of existing literature, an alternative system of street classification should be formulated, to take account of the street as a total entity. A summary of the types of changes

Table 5.1: Summary of potential changes to national highway guidance

Document	Type of guide	Nature of changes proposed
Design Manual for Roads and Bridges (DMRB)	Highway design guidance and specification for national trunk roads. Includes some specific guidance for urban roads.	Requires greater balance in the geometric and layout presumptions for the design of trunk roads (or other principal routes) in urban areas. Needs to recognise the greater requirement for pedestrian access across and along the street. Express clearly whether and how these guides can be used on local roads and what discretion can be used in their interpretation.
Design Bulletin 32 (2nd Edition), *Residential Roads and Footpaths*	National design guide for the layout, size and specification for residential roads. (Supplemented by *Places, Streets and Movement*, a companion guide.)	The guidance presumes new streets are designed around the vehicle flows rather than the vehicular flow being designed to suit the urban form and land use activities. Change guidance to reflect the more shared nature of streets, to emphasise the use and adoption of home zones, and to allow different design specifications within the street length, dependent on the adjoining activities.
Local Transport Notes	National guidance on transport and highway design for local roads for cyclists, buses, traffic signals and signalised crossings.	Particularly the advice on pedestrian crossings could be revised to address the crossing as a necessary and important infrastructural element for pedestrians. Crossing points should be selected more inclusively with pedestrians/community groups, not based solely on traffic requirements.
Traffic Advisory Leaflets	Series of advice leaflets based around traffic calming, cycling and bus measures.	A series of revisions that need to bring this useful advice up to date and make it more exemplary in urban design and movement terms. Greater understanding of the relationship between urban form, building line and movement routes across roads needs to be conveyed, as highlighted in *Places, Streets and Movement*. Streetscape effects need to be highlighted, as is done in the Historic Core Zone examples, and better illustration of integrating traffic management structures into townscape is required.
Highway Code	Road user behaviour guide aimed at all road users but primarily read by motorists for passing a driving test.	Revise to address the shared nature of streets in urban areas. Need to place greater emphasis on driver behaviour to pedestrians.

proposed is shown in Table 5.1. A detailed list of potential changes to key highway guidance is shown in Appendix 5.

The implementation of this proposal might be phased in three stages:

- **Stage 1** — advice to clarify the use of existing documents

- **Stage 2** — revision of existing guidance literature, as indicated in Table 5.1 and Appendix 5

- **Stage 3** — preparation of detailed guidance in the design of local roads in urban areas (see Section 2.3)

5.2 Street design and management

> *Recommendation 4: Local authorities should introduce cross-sectoral management control for the administration of streets, with the aim of establishing an integrated approach to the public realm.*

Currently, most local authorities control and manage streets through a variety of departments, in particular split between planning, environmental and highway functions, the latter often sub-dividing street maintenance from traffic and highway design. This division of labour reinforces the barrier between the different professions and skills responsible for the street, with the highways department often assuming the dominant role. The paramount need is for a coordinated cross-sectoral approach that treats the street as a single entity, rather than a series of separate spheres. This includes the establishment of clear shared planning and highway aims and objectives, the sharing of information and the systematic review of completed schemes. Procedures for developing this approach should be modelled and built into quality management systems by the local authorities with jurisdiction over the public realm, ensuring cross checks and joint approaches are used.

The exact form of cross-sectoral management will vary depending on the size and type of authority and will vary when there is a two-tier authority involved. Whatever arrangement is introduced, it will benefit from having the involvement and support of senior councillors. The appointment of a champion for streetscape matters, analogous to the idea of a local authority design champion, will help focus attention on the street as the key element in the public realm.

> *Recommendation 5: In urban areas, the District Council or other local authority should have responsibility for all roads and streets (other than motorways).*

This recommendation is aimed at removing the anomaly of two-tier authorities (usually a District Council and County Council) having responsibility for different classes of street in the same town, and the discordance of design and management which follows from this division of responsibility. Unitary authorities have greater understandings of local issues, including land use planning and how they affect street design. This will link highway, planning and community strategies closely and accord with the joined up thinking approach promulgated by the Planning Green Paper. Strategic 'traffic' authorities will still have a place in major metropolitan areas but their role must be limited to traffic speed, queue lengths, congestion charging, etc., not to physical highway design.

The implementation of this recommendation can be achieved by strengthening highway agency agreements between County Councils and District Councils and other local authorities. Agency agreements will emphasise the need to protect and enhance the mixed nature of the users of streets.

5.3 Design detail

> *Recommendation 6: Urban design should form a key component of all training courses for professionals dealing with streetscape and highways, including the proposed inter-professional certificate in urban design.*

There are two ways of improving the performance of professionals dealing with street design, one dealing with improved holistic urban design skills, and the other with broadening the understanding of technical issues surrounding highway regulation and design. This recommendation urges that all training in the built environment and transportation engineering contain units on urban design. This would include streetscape, highway regulation and design, and be available not only for Continuing Professional Development (CPD) but also at all training levels from National Vocational Qualifications, through Ordinary and Higher National Certificates and Diplomas, to undergraduate and post-graduate courses.

The advantage to streetscape design of such a system of training and qualification is potentially enormous, especially if it is taken up by professions such as highway engineers and enforcement officers whose role in the long-term care of the street is crucial. Equally, it needs to be emphasised that urban designers and planners should be aware of the technical aspects of street design, so that they are fully versed in the complex roles that the typical street has to play. In other words, to achieve an integrated approach to streetscape design, all those involved must seek to at least understand, if not profess, the full range of appropriate skills.

CABE Urban Design Skills Working Group (CABE, 2001) has proposed the idea of an inter-professional qualification in urban design, to be awarded to members of appropriate institutions who complete a recognised course of study. It has stressed that in order for CPD in Urban Design to be made attractive and effective to all professions, it must offer tangible benefits, both in the form of acquired knowledge and skills and through the award of a recognised qualification. All urban design is about shaping and managing the public realm, which includes the street, and all such training courses should contain inter-discipline components on streetscape design and street regulation.

> **Recommendation 7: Development Plans (or their future successor) and Local Transport Plans should include specific strategies aimed at the improvement and maintenance of streetscape.**

As currently constituted, Development Plans and Local Transport Plans (LTPs) deal with almost every aspect of the urban environment and transport, but omit specific policies for the maintenance and improvement of streetscape. Buildings and other uses along the street are included, plus the modes of transport which rely on the street, but the street itself is largely ignored. This is surprising especially since the objectives of LTPs include the protection of the environment, the promotion of integrated transport and the improvement of pedestrian comfort and convenience. The inclusion in community strategies of public realm policies to maintain and enhance the local streetscape will help express aims and objectives and focus attention on the design standards appropriate to the locality. The distinctive characteristics of streetscape in different parts of the country can be highlighted. Key performance indicators can be set out to measure the benefits of these policies over time and to express their relationship to each other and to community strategies.

If, as suggested in the Green Paper on Planning (DTLR 2001c), Unitary Development Plans and Structure Plans are replaced by Local Development Frameworks, there will still be an opportunity to identify local streetscape policies. The more detailed local action plans, focused on specific areas of change, can address the design of new streets and improvements to existing ones, identifying proposals which may be the subject of developer contributions. Strong vernacular characteristics of streets should be identified and policies identified for their conservation and enhancement.

> *Recommendation 8: Traffic Sign regulations and guidance should be revised so that local authorities can vary the size of signs according to local context as well as vehicle speed.*

In an ideal world, roads and streets would require no signage, but the needs of road safety and direction-finding mean that signs of some kind are essential. The aim of this proposal is not to replace the existing range of road and street signs; rather it is intended to introduce some flexibility in the use of such signs, to allow more thought to be given to the impact of signage on the local environment. The clarity and logic of signage in some European cities contrasts strongly with most English experience. Guidance suggests sign design is based on driver speed, so measures to reduce vehicle speed to allow signage (especially direction signage) to be smaller should be considered. The size of multiple signs should also be carefully considered to reduce the impact of large map and advance direction signs on built up areas. Steps to alter driver behaviour by unsigned junctions, as introduced in Holland, should also be looked at as ways to reduce street clutter.

5.4 Licensed operators

> *Recommendation 9: The principles of 'lane rental' and 'overcharging' systems for utility works in the street are supported and should be extended nationally if pilot projects increase street quality and reduce disruption.*

The powers granted to utility companies to make openings in the street are the single greatest cause of annoyance to street users both because of the disturbance while projects are in progress and because of the low quality of reinstatement works. It would appear from the case studies that the low level of charging available for staged local authority inspection of streetworks, plus the low financial penalties for inadequate reinstatement, allow utility companies to perpetuate working practices that do much to detract from the quality of the streetscape. As one of our consultees put it 'The utilities are running amok in our streets'. European cities exercise much tighter control over works by utilities (see the example of Freiburg discussed in Appendix 3). The current review of some Regulations under the New Roads and Street Works Act, being carried out by the DTLR, will ideally strengthen local authorities' powers of supervision of utility street works.

The current level of charges on utilities for inspections is based on only a third of the number of works done. This is insufficient to recompense authorities for the staff time and costs to carry out appropriate inspection. Raising charges would be a relatively small cost for utilities compared to the current major public cost in poor highway reinstatement. Local authorities should have powers to recover costs associated with inspecting all statutory undertakers' works. These are not available under current legislation.

Powers to fine utilities exist. These need to be set at a level which will pay for the local authority staff and systems costs of managing the final reinstatement by others of any poor street works carried out by utilities. When they see that fining works, they will be prepared to give priority to this. Some local authorities could team up to share such performance monitoring and fine collection systems.

The time charging for occupation of the road or the charging for overstaying beyond an agreed programme both generate potential income streams to maintain monitoring systems. Other 'innovations', such as micro-tunnelling (popular in Germany) or the use of combined service ducts, could be developed by the Highways Authority Utility Committee, the National Joint Utilities Group and CARD (Campaign Against Road Disruption).

> *Recommendation 10: The Control of Advertising Regulations should be applied more considerately in streets themselves as well as to development fronting streets.*

Although outdoor advertising has not been highlighted as a major issue in our case studies, there are a number of issues relating to advertising that have been raised by the project consultees (see Section 4.4). Areas of advertisement control usually coincide with Conservation Areas or areas of 'special control for advertising' when not covered by the deemed consent regime of the Advertising Regulations. Planning authorities are, on the whole, aware of advertisement impact and are sensitive to the controls required, even to advertising on street furniture. Efforts should be extended beyond the boundaries of advertisement controlled areas to protect the settings of such areas. The use of the deemed consent regime to extend advertising to street furniture has been under review by the DTLR. The use of shroud advertising on buildings is equally contentious and needs to be clearly dealt with under the regulations.

Two other issues have been highlighted in discussions with consultees:

- The siting of above ground equipment, such as junction boxes and telecommunication boxes. It has been suggested that to encourage their relocation below ground they should be the subject of rental agreements with local authorities.

- The possibility of conflicts of interest between the control of advertising by local authorities in their role as planning authority and the revenue-raising opportunities which advertising on street furniture presents.

These are two aspects of streetscape management that merit further research.

5.5 Long-term care of the street

> Recommendation 11: Involving the local community in the care of the streetscape should be encouraged through promoting local community trusts for the improvement and management of streets.

Although streets are the key element in the public realm, the public as a whole feel little sense of direct responsibility for them. Ultimately, the long-term quality of our streets depends on the degree of public commitment to their management and care.

That commitment finds expression through the work of local councils, but at the level of the individual street or group of streets there may be issues which are best dealt with through direct community involvement; a partnership between residents, businesses, community groups, etc., with the local authority to make things happen in a particular area. A model for such action exists in the Safe Routes to School Programme, which allows schools to engender proposals for streetscape and transport improvements, to be carried out in partnership with local councils. The Local Government Act 2000 encourages such initiatives through setting as an objective for local authorities 'the promotion or improvement of the environmental well-being of their area'.

The mechanism for community-based care of the street might be a local trust, technically assisted and seed-funded through a central agency, such as the Civic Trust. The remit for the local trusts would be drawn wide to allow a variety of projects and initiatives. They may include Business Improvement District Principles, raising additional local rates, or they may use other funding opportunities based in the community, like Local Strategic Partnerships. In all cases they will be locally focussed on design and appearance, economic improvement, promoting and organising the partnership's approaches.

5.6 The street as shared space

> *Recommendation 12: The Highway Code should be rewritten to place greater emphasis on the multiple use of streets, rather than mainly vehicle movement.*

The quality of our streets is not just a matter of their design and layout — it also depends on how they are perceived and used. They are a shared space which can only function well when everyone recognises the ground rules for coexistence within them.

Pedestrians have first claim to attention as users of the street because almost everyone makes a local journey on foot at some time in the day, and for many people (especially parents and young children) walking is the main form of transport. The comfort and convenience of pedestrians — and the encouragement of more people to walk — depends on the detailed quality of the local environment. It also depends on an understanding of the give and take between the different users of the street. The needs of the disabled are particularly important in this respect.

At present, the only guide to the street which most people know is the *Highway Code* which, like highway literature in general, is mainly about the vehicular use of roads and streets. Although recently revised, it makes insufficient distinction between motorways, trunk roads and principal roads, where vehicle use predominates, and local roads and streets whose main characteristic is their shared use by pedestrians, cyclists and vehicles. Also, it is addressed mainly at drivers (or people learning to drive) and says little about what are the rights and duties of pedestrians. Pedestrians are confused about what they can and cannot do, which only increases their sense of vulnerability on the street.

The junior version of the *Highway Code*, called *Arrive Alive* (DETR, 2000), could usefully serve as a model for revising the main document. Because it is aimed at people too young to drive, it opens with sections on walking and cycling, thus stressing uses of the street which normally are treated as only secondary. In the revised version of the *Highway Code*, without the loss of detailed advice to drivers, a similar stress should be placed on other users of local roads and streets, and on how all forms of movement can be made to coexist. Similarly, a drive to promote its use by pedestrians and cyclists would steer long-term appreciation of how shared use of the street works.

5.7 Implementation and the order of priorities

The evidence of the case studies shows that the improvement of our streetscapes is beginning to happen, but it still has a long way to go. The question is, what will further the achievement of better quality projects over the next few years?

Almost all of the recommendations in this report are capable of being embarked upon in the immediate future — some, such as Recommendation 6 on training, or Recommendations 9 and 10 on utilities streetworks, are already being actively considered. None of the recommendations require significant changes to primary legislation, which might be a major cause of delay.

The possible time-scale for implementing the recommendations, and the lead responsibility for carrying them through, are summarised in Table 5.2. This also shows the scale of resource implication for each proposal.

Table 5.2: Implementing the recommendations

*** Time Scale**

Short term = 1–2 years

Medium term = 2–3 years

Long term = 3 years or more

Recommendation	Lead responsibility	Time-scale*	Resource implications
1 Summary guide	DTLR	Short term: Early win	low
2 Design decision audit trail	Highway Authorities	Short term	medium
3 Review of guidance literature	DTLR	Long term	medium
4 Cross-sectoral management of streets	Local authorities	Medium term	low
5 Responsibility for urban roads	DTLR	Medium term	low
6 Inter-professional design training	Professional institutions/ UDAL	Short term	medium
7 Development plans/LTPs	Local authorities	Long term	medium
8 Traffic signs	DTLR	Medium term	low
9 Utility lane rental and overcharging	DTLR/local authorities	Medium term	high
10 Control of advertising	DTLR	Long term	low
11 Long-term care	DTLR/community organisations	Long term	medium
12 *Highway Code*	DTLR	Short term	low

The evidence gathered suggests that the most urgent proposals put forward here are:

- the production of a summary guide to existing statutes, regulations and design guidance (Recommendation 1)

- the introduction of audit trails by highway authorities as a record of decision making in the design and implementation of streetscape works (Recommendation 2)

- the introduction of streetscape design and street works to further education technical and professional built environment courses, and to an inter-professional qualification in urban design (Recommendation 6)

■   the reform of the charging regime relating to utilities streetworks
     (Recommendation 9)

The responsibility for implementing these measures is spread across a number of
organisations and government bodies.

The recommendation for the rewriting of the *Highway Code* does not, strictly
speaking, derive from the evidence of the case studies. But it does relate to the
increasing general concern about conflicts over the use of the street, and the
vulnerability of pedestrians. The announcement that this revision has been put in
hand would be a clear sign of a commitment to this aspect of the public realm.

Generally speaking, most of the recommendations are intended to reinforce policies
that already exist, but which have not been as effective as they should be, or to
enhance the powers which are already available to local authorities. What is needed
is not a wholly new regime but the invigoration of the systems and ideals that
already exist.

Summary

Recommendation 1

A succinct summary should be published, detailing the statutes, regulations and design guidance affecting streetscape design and management. This summary should distinguish clearly between advisory and mandatory documents.

Recommendation 2

Highway authorities should, under Best Value, establish an audit trail for design decisions affecting the streetscape, to show how design guidance, people's needs and vehicle movements have been accommodated.

Recommendation 3

Existing guidance on highway design should be rewritten to bring it in line with Government policy on design, sustainability and the urban realm.

Recommendation 4

Local authorities should introduce cross-sectoral management control for the administration of streets, with the aim of establishing an integrated approach to the public realm.

Recommendation 5

In urban areas, the District Council or other local authority should have responsibility for all roads and streets (other than motorways).

Recommendation 6

Urban design should form a key component of all training courses for professionals dealing with streetscape and highways, including the proposed inter-professional certificate in urban design.

Recommendation 7

Development Plans (or their future successor) and Local Transport Plans should include specific strategies aimed at the improvement and maintenance of streetscape.

Recommendation 8

Traffic Sign regulations and guidance should be revised so that local authorities can vary the size of signs according to local context as well as vehicle speed.

Recommendation 9

The principles of 'lane rental' and 'overcharging' systems for utility works in the street are supported and should be extended nationally if pilot projects increase street quality and reduce disruption.

Recommendation 10

The Control of Advertising Regulations should be applied more considerately in streets themselves as well as to development fronting streets.

Recommendation 11

Involving the local community in the care of the streetscape should be encouraged through promoting local community trusts for the improvement and management of streets.

Recommendation 12

The Highway Code should be rewritten to place greater emphasis on the multiple use of streets, rather than mainly vehicle movement.

Conclusion

6.0

6.0   Conclusion

Except at times of crisis, we take our streets for granted. We use them relentlessly, and are furious when they let us down because of roadworks or congestion; we regard them as a utilitarian necessity, rather than as a positive pleasure.

But there is now a growing realisation that we cannot treat streets in such a casual manner, partly because they play a crucial role in our economic and social well-being and also because they are at the heart of what we look for in our local environment. The aim of creating a more sustainable society based on the husbanding of our resources (especially resources for transport) depends on the quality of our streets. This means that conflicts over the use of the street have to be given a much greater priority.

These are matters which the design of quality streetscape can address. Here, as in other aspects of the urban environment, design can make all the difference to people's everyday experience, and can influence how they choose to travel. A well-designed street is one which it is a pleasure to walk down and share with other users.

What this study has shown is that efforts are being made throughout the country to improve our streets. But the success of these efforts has been limited, for four main reasons:

- The primacy still given to vehicle movement as the main element in street design

- The lack of a clear focus in local authorities for the design and management of streets

- A shortage of design expertise in the various professions that have responsibility for streets

- The failure of utilities to acknowledge their role in maintaining quality streetscapes

Of the twelve case studies analysed in the preparation of this report, the two Historic Core Zone projects (in Bury St Edmunds and Shrewsbury) have been most successful in achieving a high-quality result. There is no fundamental reason why the care that is taken for the historic areas of our towns should not be applied elsewhere. Indeed, because many other areas start with fewer assets, they deserve all the more care.

The recommendations of this report address the design and management process, rather than specific design solutions. They focus on the literature of highway design, the structure of local government street management and the role of the utilities. They also consider the underlying issues of training and the public perceptions of how the street is used.

What needs to be done is not revolutionary. It is more a need to accelerate a process of change that has begun tentatively, but which still has far to go.

References

DETR (2001). *Local Strategic Partnerships*. Government Guidance, March.

DETR (2001). *Proposed Revision to the Traffic Signs Regulations and General Directions 1994*. Consultation Paper, 5 September.

DTLR (2001). *Accreditation Guidance for Local Strategic Partnerships*. Neighbourhood Renewal Unit, October.

DTLR (2001). *Planning Policy Guidance, PPG13 Transport*.

Government response to the Environment Transport and Regional Affairs Committee's report on 'Walking in Towns and Cities', 2001.

House of Commons Environment Transport and Regional Affairs Committee (2001). *Walking in Towns and Cities*. HC 167-1, 30 June.

National Main Street Centre (2001). http://www.mainstreet.org (accessed November 2001).

Urban Forum and Community Development Foundation (2001). *The LSP Guide*.

Urban Land Institute (2001). *Residential Streets*, third edition. Washington D.C.

DETR (2000). *A good practice guide for the development of Local Transport Plans*. April.

DETR (2000). *Guidance on Full Local Transport Plans*. March.

English Heritage (2000). *Streets for All*.

English Partnerships (2000). *Urban Design Compendium*.

IHT (2000). *Guidelines for providing for journeys on foot*.

UDAL and ICE (2000). *Designing Streets for People*.

UDAL and ICE (2000). *Returning Roads to Residents*.

(1999). *Highway Code*. The Stationery Office.

DETR (1998). *Walking in Great Britain*. Transport Statistics Report, HMSO.

IHT (1997). *Transport in the Urban Environment*.

Institution of Transportation Engineers (1997). *Traditional Neighbourhood Development Street Design Guidelines*. ITE Transporation Planning Council Committee 5P-8, , Washington.

Royal Fine Art Commission (1997). *Improving Design in the High Street*. Architectural Press.

Sauvain, S. J. (1997). *Highway Law*. Sweet and Maxwell.

Civic Trust /English Historic Towns Forum (1994). *Traffic in Townscape: Ideas from Europe*.

English Historic Towns Forum (1994). *Traffic in Historic Town Centres*.

Civic Trust /English Historic Towns Forum (1993). *Traffic Measures in Historic Towns*.

DTLR (1991). Proposed revisions to regulations in the New Roads and Streetworks Act 1991 — various Regulations. Consultation Papers, see http://www.detr.gov.uk/.

Department of the Environment (1987). *Roads and Traffic in Urban Areas*.

McCluskey. J. (1979). *Road Form and Townscape*.

Ministry of Transport (1966). *Roads in Urban Areas*.

Buchanan, C. (1963). *Traffic in Towns*.

Nairn, I. (1956). Casebook — Roads. *Architectural Review*, December.

Nairn, I. (1956). *Outrage*. Architectural Press.

Sharp, T., Gibberd, F. and Holford, W. G. (1953). *Design in Town and Village*. Ministry of Housing and Local Government, HMSO.

Alker Tripp, H. (1942). *Town Planning and Road Traffic*. Edward Arnold, London.

Eno, W. P. (1939). *The Story of Highway Traffic Control 1899–1939*. The Eno Foundation for Highway Traffic Control.

Statutes

Local Government Act 2000.

Town and Country Planning (General Permitted Development) Order 1995, Statutory Instrument 1995 No. 418.

Traffic Signs Regulations and General Directions SI 1994 No. 1519.

New Roads and Streetworks Act 1991.

Road Traffic Act 1991.

Streetworks (Sharing of Costs) Regulations 1991.

Environmental Protection Act 1990.

Road Traffic Act 1988.

Road Traffic Regulation Act 1984.

Highways Act 1980.

Design guidance

DoT/DoE (1994). *DB32 Residential Roads and Footpaths*. HMSO.

DETR (1995). *Local Transport Note 1/95, 2/95 (Pedestrian crossings)*.

Design Manual for Roads and Bridges (1994, as current at August 2001)

- TA 23/81 *Junctions and Access: Determination of size of roundabouts and major/minor junctions*
- TD 36/93 *Subways for pedestrians and pedal cyclists*
- TD 42/95 *Geometric design of major/minor priority junctions*

Traffic Advisory Leaflets

- TAL 1/87 *Measures to control traffic for the benefit of residents, pedestrians and cyclists*
- TAL 7/93 *Traffic Calming Regulations*
- TAL 12/93 *Overrun Areas*
- TAL 13/93 *Gateways*
- TAL 4/94 *Speed Cushions*
- TAL 9/94 *Horizontal deflections*
- TAL 7/95 *Traffic islands for speed control*
- TAL 12/97 *Chicane Schemes*
- TAL 9/99 *20mph speed limits and zones*

Transport for London Guidance

- *Walking Schemes: Good Practice Guidance for London*

Appendix 1.0

Case studies

Coldharbour Way
Aylesbury

CASE STUDY

Street type	New suburban distributor
Intervention	New street
Principal authority	Buckinghamshire County Council

PROJECT RESPONSIBILITIES

Highway authority	Buckinghamshire County Council
Design	John Simpson & Partners — masterplan Barton Willmore Partnership — detail
Project management	McLean Homes, Wimpey Homes, Bryant Homes, Taywood Homes
Maintenance	Buckinghamshire County Council

DESCRIPTION

- Fairford Leys is a new mixed-use urban extension to Aylesbury consisting of 1800 homes plus a school, church and community centre, shops and offices

- The area is largely made up of terraced buildings giving continuous built frontage to roads and public spaces

- Consistent materials have been used in the buildings and streetscape

DESIGN OBJECTIVES

- To give this new place an identity and sense of community rooted in the context of Aylesbury and its surroundings

- To reduce the dominance of the car and provide walkable quarters within the development

VISUAL APPRAISAL

A new mixed-use urban extension with much thought given to urban design and sustainable development principles.

The main route through the development has been designed to reduce traffic speeds by ensuring that straight stretches of road are short. Traffic flows and speeds are low, creating a more pedestrian feel and taking away the need for designated crossings.

Street furniture is of an historic style and is well coordinated. Paving colours are simple, although bonding of elements is sometimes inappropriately aligned to suit the vernacular style buildings. The bond of paving also alters between developer plots.

Little use of road signs removes potential clutter from the streetscene.

QUALITY ASSESSMENT

Comfort & safety	Adequate comfort for pedestrians, well overlooked. Drop kerbs have been used to existing standards.
Domination by functions	Well balanced for mixed-use area but maintains character as a street.
Visual simplicity	Materials used are simple. The lack of street planting and wide carriageways provides a bleak landscape.
Utilities subordinate	Noticeable remediation but generally not a problem given the tarmac materials used.
Fitting to character & activity	Appropriate traffic calming by layout to create activity cores where pedestrian activity can mix safely.
Ordered for access & storage	Parking in designated bays or squares. Some parking behind buildings in mews courts.

Top to bottom:

The Local Highway Authority's requirements have meant that the street is now characterised by the sinuous curves of a standard carriageway width

There is little planting in the street, which gives the area a cold bleak feel. This will improve as private planting matures

Owing to adoption requirements, paving has been used to mark out a standard carriageway even in shared surface areas

KEY ISSUES

- Good use of layout to create a series of places that help to calm traffic speeds and focus activity

- Original concept has been compromised by Highway Authority's insistence on a standard carriageway detailing on internal roads and the spine road. Use of the variable widths outlined in the masterplan has therefore been prevented

- More on street planting might have helped to soften the street scene and add character to the development

- Although simple and well coordinated, the street furniture, paving and building styles combine to create an uneasy mix of modern and vernacular — leading to a confused visual effect

Ladypool Road
Birmingham

CASE STUDY

Street type	Suburban main street
Intervention	Environmental improvements
Principal authority	Birmingham City Council

PROJECT RESPONSIBILITIES

Highway authority	Birmingham City Council
Design	De Montford University Birmingham City Council
Project management	Birmingham City Council
Maintenance	Birmingham City Council

DESCRIPTION

- A busy through route in a suburban area of Birmingham

- The street is in a lively shopping and leisure area

- The improvements, made as part of an SRB funded scheme, attempt to recognise the high proportion of local residents from ethnic minorities and reinforce their sense of identity with the area

- Improvements to the street mainly consist of new street furniture, traffic calming elements and some improved pedestrian facilities

- The improvements complement a shop enhancement programme that has already been completed

DESIGN OBJECTIVES

- Reinforce the identity of the area through distinctive street furniture

- Improve the pedestrian environment

- Manage traffic levels and speeds passing through this busy area

Top to bottom:

The street is a busy shopping and leisure area in an inner suburb of Birmingham

Shops spill-out onto the street with stalls and rails of goods in some areas

Differing priority measures are used to control traffic speeds, although roundabouts do not give pedestrians the opportunity to cross

Drop kerbs have been provided at signal controlled crossings

VISUAL APPRAISAL

Dropped kerbs and tactile paving have been used to improve pedestrian crossings. New street furniture in a distinctive 'ethnic style' has been installed along the length of the street.

The bright green and yellow street furniture is visually obtrusive and fails to complement the land uses that line the street in the way that was intended. Paving bonds are spoiled by pattern making in blocks.

A busy pedestrian and vehicular route, the street has the appearance of being 'full' with little space in the carriageway due to parked vehicles. Too much carriageway space at some junctions.

A certain amount of clutter was witnessed from shop fronts extended into the street, however this adds vibrancy and contributes to the colour and character of the area.

QUALITY ASSESSMENT

Comfort & safety	The street is well overlooked in most areas and footway widths are reasonable. Traffic volumes are too high, reducing the feeling of safety.
Domination by functions	High vehicle speeds and volumes, as well as illegal parking, mean that the street is car dominated.
Visual simplicity	The street is not cluttered with signage or too much street furniture. The street furniture is visually obtrusive.
Utilities subordinate	Does not appear to be a problem.
Fitting to character	Dominated by the through movement of traffic. Materials used are appropriate for the location. Street furniture's 'ethnicity' is questionable.
Ordered for access	Illegal parking takes place along the length of the street. Shop overspill onto footway needs monitoring.

KEY ISSUES

- Top down version of ethnic identity or distinctiveness is not always successful in generating a sense of identity

- Parking control to enforce current restrictions

- Traffic calming events could be combined with pedestrian crossings to engender a sense of pedestrian priority

- The green street furniture is not an easy colour to fit into the streetscene and is likely to weather poorly

Top to bottom:

The street has poor enclosure in some areas where the terraced building style breaks down

Illegal on-street parking and high traffic volumes mean that cars dominate the street despite reasonable footway widths

The street furniture is designed to reinforce the ethnic character of the street although the design quality is crude

Top to bottom:

The street forms part of the planned medieval centre of Bury St Edmunds

The pedestrian environment has been improved with wider pavements and lower traffic speeds

Sufficient parking has been provided in a simple manner right in the centre of town

The street is one-way to traffic and forms part of a traffic management system for the town

Hatter Street
Bury St Edmonds

CASE STUDY

Street type	Historic Core Zone
Intervention	Historic Core Zone — Environmental improvement and traffic management
Principal authority	St Edmundsbury Borough Council

PROJECT RESPONSIBILITIES

Highway authority	St Edmundsbury Borough Council as an agent for Suffolk County Council
Design	St Edmundsbury Borough Council
Project management	St Edmundsbury Borough Council
Maintenance	St Edmundsbury Borough Council for Suffolk County Council

DESCRIPTION

- Hatter Street is a moderately busy mixed-use street in central Bury St Edmunds

- It forms part of a planned medieval grid system and has been improved as part of an Historic Core Zone initiative

- The street is one-way for traffic with on-street parking provided down one side

- A pedestrian shopping area joins the street both at a central point and at one end

DESIGN OBJECTIVES

- Enhance the appearance of the street in a way that befits its historic character.

- Improve quality of materials in the street

- Traffic management as part of a system for the whole town centre

- Improve pedestrian environment while still providing through movement for traffic and sufficient parking levels

- Allow limited delivery/loading access

VISUAL APPRAISAL

The street is an historic urban street within the centre of town. The street houses buildings from many different periods along a strong continuous building line.

The paving and street furniture have been well coordinated, with a simple and easily understood palette of materials and colours.

The one-way system allows parking to be on-street within a narrow area. The parking is well ordered and does not detract from the visual qualities of the scheme.

Clutter levels are very low providing an attractive street scene — lighting and street signage could have been wall mounted to improve the situation still further.

QUALITY ASSESSMENT

Comfort & safety	Low traffic speeds. Well overlooked — feels safe to walk down. Drop kerbs are used and there are no difficult changes in level, the paving is also in good condition.
Domination by functions	Cars are highly visible but do not dominate. Low traffic speeds encourage pedestrians to wander, however very few cyclists were using the street during the study.
Visual simplicity	Low key — well coordinated paving and furniture. Clutter levels very low but could be reduced even more by wall mounting signs and lighting.
Utilities subordinate	A few problems with reinstatement but generally not a problem. Service access covers were coordinated with paving materials.
Fitting to character	Materials used fitting to the historic nature of the area. Provides a safe comfortable route as well as useful parking close to the main shopping area.
Ordered for access	Good parking and loading provision, which was well used. Pedestrian access points were well detailed.

KEY ISSUES

- An Urban Designer managed the project up to detailed design stage, with advice from Highway Engineers, providing a much more site-sensitive solution than might have resulted from a standard approach

- Integrated approach to parking materials and signage due to effort put into the overall Historic Core Zone scheme

- Appropriate use of materials and well coordinated street furniture

Top to bottom:

The materials used are simple, of high quality, and appropriate to their setting — square bollard base fits

Although cars line the street they do not dominate the space

Street furniture could have been wall mounted to reduce clutter still further — rectangular paving

Cowcross Street
London

CASE STUDY

Street type	Inner urban local street
Intervention	Environmental improvement
Principal authority	LB Islington

PROJECT RESPONSIBILITIES

Highway authority	LB Islington
Design	Alan Baxter & Associates
Project management	Alan Baxter & Associates for street frontagers
Maintenance	LB Islington

DESCRIPTION

- It is a busy route from Farringdon Underground and ThamesLink station to the City of London

- The street has a variety of building types and uses, ensuring a good level of activity at all periods of the day

- The street is one-way for vehicular traffic

- A conservation-led scheme has sought to improve the environment in the street, particularly for pedestrians

- SRB funding was secured for 50% of the costs of the scheme, with the remainder covered by property owners' contributions

DESIGN OBJECTIVES

- Improve the public realm quality in the area

- Enhance the appearance of an historic street and the setting of buildings within a conservation area

- Improve the feeling of safety for pedestrians using this busy route

- Provide ordered access for deliveries and storage of parked cars

Top to bottom:

Cowcross Street is a busy pedestrian route to the City of London

Pavement cafés provide an active frontage to the street but can infringe on the pedestrian area

No designated cycle facilities are provided and cyclists often ride illegally against the flow of traffic

Wide pavements provide a safe and comfortable environment in which to walk or interact

VISUAL APPRAISAL

Good sense of enclosure. Buildings vary in height and orientation to the street, ensuring visual interest. Although shaded, sunlight breaks through the varied roofline, changing the appearance of the street at different times of the day.

Simple materials have been used, which are appropriate for the urban setting.

Although there is little planting, there is enough to soften the hard edges of the street at certain points.

Street furniture is well coordinated and clutter is low. Illegal car parking tends to spoil the overall look by filling up the street with vehicles.

QUALITY ASSESSMENT

Comfort & safety	The potential benefit of the wide footways is let down by cracked paving and litter. The street is active and overlooked.
Domination by functions	Traffic speeds can be high on the straight section due to the one-way nature of the street but generally a balance is achieved.
Visual simplicity	Simple materials and coordinated street furniture have contributed to a visually simple streetscene. Lighting and signage are wall mounted, reducing clutter.
Utilities subordinate	Services restricted location of bollards to prevent footway parking. Services have been dug up in many places however, and reinstatement has generally been poor.
Fitting to character	The materials chosen suit the historic nature of the site. Simplicity of layout and colours gives a suitable backdrop to the varied buildings and busy street life.
Ordered for access	Parking in bays but many people park illegally on yellow lines. Loading takes place all along the street rather than in designated bays. Vehicles clutter the street in places.

KEY ISSUES

- Poor utilities reinstatement and maintenance spoil the look of the scheme
- Cracking of paving materials from vehicular overrun following poor reinstatement
- Too much clutter from illegally parked vehicles
- Crossovers vary along the street with pedestrians forced to make a change in level in some cases
- Control of A-boards and tables and chairs is required as these tend to clutter the pedestrian environment (though also prevent footway parking)

Top to bottom:

Wall mounted street furniture has been used to reduce clutter levels

Utilities interventions are numerous and almost all have been poorly reinstated

Maintenance of the street has not been effective since the improvements

West Street
Dunstable

CASE STUDY

Street type	Trunk road through rural town centre
Intervention	Traffic management
Principal authority	DTLR/Highway Agency (HA)

PROJECT RESPONSIBILITIES

Highway authority	DTLR/HA
Design	Carillion URS for HA
Project management	Carillion URS for HA
Maintenance	Carillion URS for HA

DESCRIPTION

- A dual carriageway regional road passing through a busy rural town centre

- Improvements have been made to the street at various stages

- Bus lanes have been added to ease their movement through the centre

- The town centre is almost completely severed by the busy carriageways — crossings are difficult, especially at junctions

DESIGN OBJECTIVES

- Improve facilities for buses within the town centre

- Protect the pedestrian area from high traffic levels

- Improve paving and street furniture quality

Top to bottom:

An historic through route, which once gave life to the town, now severs it due to high traffic levels

Clutter from signage and bollards is particularly obtrusive

Coloured lanes have been used to delineate bus lanes in the centre

Guardrails have been used along most of the street to prevent pedestrians crossing outside of designated areas

VISUAL APPRAISAL

The street is a busy highway, with high-speed traffic, which dominates the visual appearance of the town.

There are major problems with clutter and guardrailing near a dangerous junction where two major trunk roads intersect at the heart of the town centre.

The street is an important shopping area and, as such, has a busy street environment.

Street furniture and paving has not been coordinated in the improvement scheme.

A layering effect has occurred where previous schemes are added to over the years, leaving a legacy of varying furniture styles and colours.

QUALITY ASSESSMENT

Comfort & safety	Conflict between high traffic volume and town centre function makes for a poor quality environment for pedestrians and reduced safety.
Domination by functions	The street's function as a major trunk road is entirely dominant over other aspects of usage.
Visual simplicity	Cluttered due to various street uses and conflicts. This is increased by the continual addition of furniture without considering the existing fixtures.
Utilities subordinate	Does not appear to be a problem.
Fitting to character	The heavy traffic flow and street furniture associated with the trunk road are at odds with the street's function as part of an important town centre.
Ordered for access	Delivery made in front of retail units, on the highway, causing a nuisance to traffic and pedestrians.

Top to bottom:

Paving bonds are irregular and junction boxes are visually obtrusive and poorly positioned

While wide pavements provide a comfortable walking area, the presence of guardrailing constantly reminds the pedestrian of the dominance of the road

An accumulation of street furniture from various periods and of various colours line the street leading to visual chaos in spite of the coordination scheme

KEY ISSUES

- The traffic role of the trunk road passing through the town is completely dominant over other functions

- An agreed materials and street furniture palette is being used for phased improvement but use of furniture is excessive and positioning is often poor

- Guardrailing has been relied upon to provide pedestrian safety rather than treating the nature of the road

- The treatment of the trunk road is specified by a contract area spanning various places and administrations in town and country rather than the qualities of the areas it passes through

Belle Isle Road
Leeds

CASE STUDY

Street type	Suburban distributor
Intervention	Traffic management
Principal authority	Leeds City Council

PROJECT RESPONSIBILITIES

Highway authority	Leeds City Council
Design	Leeds City Council
Project management	Leeds City Council
Maintenance	Leeds City Council

DESCRIPTION

- A dual carriageway residential collector road for an inter-war council estate that runs towards Leeds city centre

- The area is the subject of a housing and environmental improvement scheme

- The area is designated as a potential corridor for the proposed Leeds Supertram towards the south of the city

- Improvements to the street consist mainly of traffic calming elements and some improved pedestrian facilities

- No on-street parking is provided

DESIGN OBJECTIVES

- Improve pedestrian safety by reducing traffic capacity and speeds, and improving parking and pedestrian crossings

- Part of the ongoing maintenance of the street

- To complement the housing improvements scheme being carried out on the estate without prejudicing the potential tram route investment

Top to bottom:

Dual carriageway estate distributor built in the 1940s. Low budget lighting, signage and traffic calming

Pedestrian space quality is not as well improved as that for parked cars

Low budget but appropriate paving

Some crossings not improved at all — bus shelters from another era

VISUAL APPRAISAL

A collector road designed primarily with the motorist in mind — typical of its construction period.

The current road surface is of black and red tarmac, with grass verges and black tarmac footways.

The street is lined with low-rise council housing and feature/retaining walls in some areas.

Dropped kerbs and tactile paving have been installed to improve pedestrian crossings. Small road humps have also been used at these crossings to slow down approaching traffic.

QUALITY ASSESSMENT

Comfort & safety	Crossings are improved and traffic speeds lowered but the pedestrian environment still feels unsafe. Little overlooking in some areas.
Domination by functions	Although parking lanes have been added, reducing the carriageway width, the scale and treatment of the road is inappropriate for its residential setting.
Visual simplicity	Materials used are simple. The lack of planting in the central reserve and wide carriageways provides a bleak landscape.
Utilities subordinate	Noticeable remediation but generally not a problem given the tarmac materials used.
Fitting to character	Dominated by its function as a road. Signage and treatments used are unsympathetic to the residential nature of the street.
Ordered for access	Parking in-curtilage and some parking lanes. Pedestrian access is a problem.

Top to bottom:

Crude stand-alone build outs with excessive signage and road markings inappropriate to design speeds

Traffic dominated junctions with crude guidance markings and signage. Pedestrians unnecessarily herded with guardrails

No guardrail and simple paving. Speed cushions to prevent discomfort for buses

KEY ISSUES

- The outline plan for the 'Supertram' to use this street has blighted attempts to improve the streetscape. Planting and other improvements have been put on hold until the route is finalised

- Little has been done to reduce the impact of the car in this street. Traffic management must be coupled with environmental improvements

- The traffic calming put in place does little to benefit the pedestrian. Changes to the nature of the street have been minimal

- Excess signage urbanises the suburban green route

Barras Street
Liskeard

CASE STUDY

Street type	Rural town main street
Intervention	Environmental improvement
Principal authority	Cornwall County Council

PROJECT RESPONSIBILITIES

Highway authority	Cornwall County Council
Design	Cornwall County Council
Project management	Cornwall County Council
Maintenance	Cornwall County Council

DESCRIPTION

■ The main High Street for the town is fronted by shops, banks, library, etc., linking original market square/crossroads to other junctions to the south west

■ Mainly fronted by tall, some grand, buildings, although disfigured by set backs or petrol-filling station and gap site at the south-west end

■ Physical improvement measures include paving, furniture, planters and pedestrian crossings, although the scheme is incomplete at the junction to the south west

DESIGN OBJECTIVES

■ Improve the quality of street furniture and paving in this main shopping street for the town

■ Increase pedestrian comfort and slow through traffic

■ Remove parking from the market square

■ Use materials that will blend with the existing local architecture

Top to bottom:

The street is in a busy mixed-use area in the centre of the town

Alterations have been made to improve the comfort of pedestrians and create attractive spaces — shown here where cars once parked

Footway build-outs and bays for parking and bus stops have been added to the street

Alterations have been made to improve the comfort of pedestrians and create attractive spaces — shown here where cars once parked

VISUAL APPRAISAL

The scheme completely ignores any aspect of local vernacular streetscape. Highly distinctive wide kerbs have been enveloped by concrete paving slabs and reconstituted stone kerbs.

Well-defined spatially but detail design unsympathetic to traditional features due to complex paving bonds and use of modern materials, including coloured imprinted concrete.

Street furniture is not coordinated and pastiche historicist in style.

Any pedestrian improvements are marred by poor junctions with adjoining areas.

QUALITY ASSESSMENT

Comfort & safety	Traffic still feels very close to pedestrians. Guardrailed roundabout restricts pedestrian movement.
Domination by functions	Traffic dominates through crossings provided.
Visual simplicity	Attempts to match materials have selected busy elements. Street furniture is not coordinated and paving bonds change.
Utilities subordinate	Utilities impacts not evident except in areas of black top, which are not visually obtrusive.
Fitting to character	Attempts to provide mix of parking seating and loading but poorly detailed to maintain character of architecture. Still too car dominated for a mixed-use main street.
Ordered for access	Loading bays over capacity leading to footway parking. Planters used to infill spare space.

KEY ISSUES

- Good restoration of vehicular areas for pedestrians and provision of controlled crossings

- A strong vernacular style was evident in the area, yet the design fails to take account of this

- A detail design strategy was not followed through for the area leading to a hotchpotch of different paving bonds and street furniture

- Failure to integrate footway widening into existing scheme

- Low quality materials used over a wide area — a less ambitious scheme might have allowed for better quality

- On-street loading bays possibly under provided for

Top to bottom:

Traditional local streetscape elements, like the wide kerbs and granite slabs, shown here in an adjoining street, have not been used in the scheme

Expensive bollards used, while cheaper materials have been used in other areas

Build-outs are merely add-ons to the original footway, not integrated with them

Top to bottom:

The street takes a large volume of traffic without feeling vehicle dominated

Avenue planting and hard landscape strips create a buffer between pedestrians and traffic

Generous footways are provided along the street — although cycles share the carriageway

Pedestrian crossings are signal controlled

St James' Boulevard
Newcastle

CASE STUDY

Street type	Inner urban major distributor
Intervention	Environmental improvement and new road construction
Principal authority	Newcastle City Council

PROJECT RESPONSIBILITIES

Highway authority	Newcastle City Council
Design	Newcastle City Council
Project management	Newcastle City Council
Maintenance	Newcastle City Council

DESCRIPTION

- Two streets were combined to create a single route along and past the city centre — replacing a difficult and circuitous one-way system

- Busy pedestrian and vehicular route between Redheugh Bridge and Gallowgate in Newcastle-upon-Tyne

- Carries large volume of traffic while attempting to allow easy pedestrian movement across and along it (on route to the football stadium)

- The road crosses some of the key radial routes into Newcastle city centre from the west

DESIGN OBJECTIVES

- Simplify the road network for the city

- Provide a distributor to take heavy traffic around, instead of through, the city centre

- To open up parts of the Historic Core, which were previously carrying heavy traffic

- The distributor should not create a barrier to pedestrian movement

- Improve pedestrian and cyclist facilities

VISUAL APPRAISAL

Little sense of enclosure due to a lack of buildings lining the street.

The tree-lined style does help to create a sense of place and soften the effect of traffic to pedestrians.

The street is visually very simple, and almost completely clutter-free.

Street furniture is simple and well coordinated and high-quality paving materials have been used throughout.

The footways provide a comfortable pedestrian environment, however crossings at major junctions are tortuous and provide little sense of safety for the pedestrian.

QUALITY ASSESSMENT

Comfort & safety	Good quality materials and few difficult changes in level. Degree of separation from traffic allows for a comfortable pedestrian environment — except at crossings.
Domination by functions	Cycle lanes provided in the carriageway, which seems unnecessary and unwanted with the high traffic levels. Generally, a balance is achieved between all users.
Visual simplicity	Clutter free — visually simple to the point of emptiness. This will change with adjoining development. Overall an easily understood and attractive boulevard style street.
Utilities subordinate	No visual scars from utilities. Coordination with utilities at an early stage avoided mains in carriageways.
Fitting to character	Well detailed, successfully combining its role as a distributor with that of an urban street for the most part. Few buildings front it at present so its suitability may change with the development around it.
Ordered for access	Parking and loading all provided in side streets. Illegal parking on the street does not seem to be a problem.

Top to bottom:

Brightly coloured surfaces and cycle lane step-ins are out of character with the simplicity of the rest of the scheme

Pedestrian crossings at main junctions are tight and staggered, with pedestrians waiting in islands

The combination of two streets has required building demolition leaving little enclosure to the street until sites are developed

KEY ISSUES

- Inner urban distributor designed so as not to create a barrier to pedestrian movement (in contrast to the central motorway, which serves a similar role to the east of the city centre)

- All crossings are at grade and the whole scheme is intended to be pedestrian and cyclist friendly

- Unfortunately, the pedestrian crossings are difficult — often with pedestrian islands in the carriageway. Detailed design has relied on standard solutions instead of carrying through the concept to the last detail

- A lack of built edges to the street provides a poor sense of enclosure

Castlegate
Nottingham

Top to bottom:

Certain sections of the street have been either partially or completely pedestrianised

A comfortable pedestrian environment has been provided, although trees may have been better than planters as a means of softening the scene

Good quality materials have been used throughout the pedestrianised areas

Planting in certain areas adds to the character of the street

CASE STUDY

Street type	Inner urban main street
Intervention	Environmental improvement, traffic management and maintenance
Principal authority	Nottinghamshire County — Highways Nottingham City — Planning

PROJECT RESPONSIBILITIES

Highway authority	Nottinghamshire County Council
Design	Nottingham City Council
Project management	Nottingham City Council
Maintenance	Nottingham City Council

DESCRIPTION

- An historic main route from the centre of Nottingham to the Castle

- Seen as a key link in the tourist trail in Nottingham as it joins the Castle with the Lace Market area (one of the oldest parts of Nottingham)

- The street crosses Maid Marion Way, part of the busy inner ring road for the city. Crossing points on this road have always been a problem and an at-grade crossing was installed after much deliberation

- The street is lined with a variety of uses and building types, from a variety of architectural periods

- Funding has been provided from both the City and County budgets for the scheme

DESIGN OBJECTIVES

- Improve links between the Castle and the city centre, as part of a tourist trail

- Improve the pedestrian environment

- Retain and enhance the historic feel of the street

- Limit traffic entering the city centre

VISUAL APPRAISAL

This is a street with three separate characters — a pedestrian zone, a two-way street, and a one-way system crossing.

The pedestrian areas and two-way section both use good quality materials, with coordinated and uncluttered street furniture.

One-way system area spoils the look of the whole scheme with clutter from street furniture which is of poor quality and uncoordinated. A staggered pedestrian crossing also blocks the street both visually and physically.

QUALITY ASSESSMENT

Comfort & safety	Although on incline there are no steep changes of level. The street is busy and feels safe. The paving quality is very good.
Domination by functions	The traffic route of Maid Marian Way does dominate the street somewhat, but only in the central section. Otherwise a balance is achieved between functions.
Visual simplicity	Colours and materials are of high quality and work well together visually. In the central section they change, as does the street furniture spoiling the effect.
Utilities subordinate	Noticeable remediation of poor quality in some areas.
Fitting to character	Materials reflect the historic nature of the route yet still provide for vehicular loads to footpaths due to the urban location. Maid Marian Way's character is not fitting to an urban mixed-use area.
Ordered for access	Some ordered parking, however illegal parking and deliveries were present. Cycle lanes unnecessarily obtuse.

KEY ISSUES

- The pedestrian crossing is staggered where Maid Marion Way cuts across the street, breaking the historic link between the Castle and the commercial centre

- Services reinstatement has been of poor quality in some areas and numerous interventions have been made by utilities

- Pedestrian area vehicle controllers damaged and not repaired — they appear not to have been needed as the signage and paving change seems to be effective in preventing vehicles in the pedestrian area

- A lack of coordination between the County and City Councils has meant that no design cohesion has been achieved. Where the ring road crosses the street the character and treatment changes dramatically. This could improve now as the city is a unitary authority with highway control

Top to bottom:

The crossing is staggered, forcing pedestrians off the line of the street

Planters used to soften the guardrails only serve to draw attention to them, and reduce visibility for children

The scheme loses continuity in this one-way section next to the ring road with on footway contraflow cycle lane

Shrewsbury High Street
Shrewsbury

CASE STUDY

Street type	Historic Core Zone
Intervention	Historic Core Zone — environmental improvement and traffic management
Principal authority	Shropshire County Council

PROJECT RESPONSIBILITIES

Highway authority	Shropshire County Council
Design	Halcrow Fox
Project management	Halcrow Fox
Maintenance	Shropshire County Council

DESCRIPTION

- Integrated traffic management and environmental improvement plan to maintain and balance access to the town centre for all users

- Waiting loading and parking management

- Narrowing and paving main carriageway to 3.5 m, widening footways and providing disabled, loading and bus lay-bys

- Provision of informal courtesy crossings

DESIGN OBJECTIVES

- To reduce reliance on the car and encourage other transport, such as walking and cycling

- To preserve and enhance the character and appearance of the area

- To limit traffic volume using the street and reduce vehicle speeds

Top to bottom:

The street is a busy pedestrian and one-way traffic and bus route in the centre of Shrewsbury

Loading has been provided in on-street bays surfaced in a different material to the carriageway

The pedestrian areas are comfortable and the buildings in the street provide interest and overlooking

Paving materials used are simple and of high quality

VISUAL APPRAISAL

The street lies in the medieval core of Shrewsbury and is lined by many historic buildings. The materials used blend well with the built character, although timber bollards are not wearing well.

A number of old routes run off/across the street, proving access and points of interest along the street.

The street is in the commercial area of the town, although is not the main shopping street, which is perpendicular to it.

One-way through traffic only allowing for on-street loading down one side and bus stops maintains access and activity.

QUALITY ASSESSMENT

Comfort & safety	Drop kerbs and good quality materials. Low traffic speeds and pedestrian priority at crossings. The street is also well overlooked and feels safe.
Domination by functions	Traffic still tends to dominate despite low speeds — this is due to the misuse of loading bays for parking. Few cyclists use the street.
Visual simplicity	Very little clutter — paving materials are simple but of high quality. Street lights and most signs were wall mounted.
Utilities subordinate	Service ducts are covered with paving material. Generally not a problem.
Fitting to character	The materials and street furniture used visually suit the historic nature of the street but some have failed technically. Pedestrians are well catered for in a busy central shopping area.
Ordered for access	Provision made for bus stops, loading and disabled parking bays using different surface materials — this is ignored and all are used for parking however.

KEY ISSUES

- Retro-fitting of street furniture has damaged original paving materials

- Use of granite setts to lower traffic speeds has been compromised by the failure of the setts under loading due to design and construction problems. Stretches of carriageway now have black-top surface

- Uncertainty as to how the construction problem could be solved has led to patches of poor quality temporary reinstatement spoiling the look of the scheme

Top to bottom:

Street lighting has been wall mounted, significantly reducing clutter in the street

Placing a junction box in an existing wall highlights how services can be integrated with their surroundings

Although most paving is well detailed, here tactile studs have been retro-fitted and the carriageway materials have failed

Top to bottom:

Raised 'Kassel' kerb provides easy access to bus for the mobility impaired

A clear and comfortable waiting area is provided

Tree-lined sections with verges, installed with the housing, soften the street — although trees are small and many are past maturity

Tarmac rather than flag paving used to fill-in the transition between the new crossing and existing paving

Wellington Road North
Stockport

CASE STUDY

Street type	Suburban distributor
Intervention	Environmental improvement
Principal authority	Stockport M.B.C.

PROJECT RESPONSIBILITIES

Highway authority	Stockport M.B.C.
Design	Stockport M.B.C.
Project management	Stockport M.B.C.
Maintenance	Stockport M.B.C.

DESCRIPTION

- The street is a suburban residential stretch of the A6. It forms part of the suburban boundary between Manchester and Stockport

- Built in a floodplain left vacant during the Victorian expansion of the two towns, it was developed in the 1930s

- Although predominantly residential in nature, part of the adjoining frontage is taken up by a biscuit factory

- The street is on a busy arterial road and bus route, which has recently been improved as part of a quality bus corridor scheme. Measures include dedicated bus and cycle lanes, raised kerbs and improved bus stop areas with new shelters

- New pedestrian crossings have also been introduced, with tactile paving and audible crossing signals

DESIGN OBJECTIVES

- Improve bus stop setting and shelter
- Allow easy access to public transport for all
- Provide priority bus and cycle lanes in conjunction with traffic management
- Improve pedestrian facilities along the route

VISUAL APPRAISAL

Wide street with little sense of enclosure — carriageway tends to dominate the space. Tree-lined sections reduce the effect of traffic, although trees are small and verges are worn or damaged.

The hard urban feel of the improvements are in contrast with the overall leafy residential feel of the area.

There is a moderate level of clutter from bus and cycle lane signage. The large directional and bus/cycle lane signs used are inappropriate for this residential stretch of street. The coloured road markings are very apparent.

Bus stops are comfortable and good quality materials have been used, their detailed design has been well considered but in isolation of context. The paving bond does not match the existing bond, and the bus shelter and flag do not coordinate.

Improvements only at bus stops, the rest of the street looks tired and materials are deteriorating.

QUALITY ASSESSMENT

Comfort & safety	Cracked paving, high traffic speed and volume, and overlooked only on one side. Bus stop area clean and comfortable.
Domination by functions	Good pedestrian crossings but poor footways with numerous vehicle crossovers. Buses can tend to dominate the carriageway.
Visual simplicity	Varying styles of paving due to improvements only being at bus stops. Coloured markings and signage are visually distracting.
Utilities subordinate	Did not seem to affect the street scene.
Fitting to character	Suitable for distributor and bus route function, but does not really feel like a residential road. Coloured pavings clash.
Ordered for access	In-curtilage (hidden by planting) and side street parking. However vehicle crossovers cross mandatory bus/cycle lanes.

KEY ISSUES

■ Scheme carried out in partnership with Manchester City Council, as the route passes through Manchester to the city centre. Part of the route goes through an SRB area in Manchester and so necessary consultation slowed down the scheme design and implementation

■ Lack of being locally adopted, this bus corridor design guidance meant that a lot of resources went into exploring ideas and consulting on details in spite of research available from London and other cities. Designs went out to comment to other districts in the Manchester area while works were on site, when they should have already been agreed prior to construction

■ Improvements only at bus stops and crossings, creating an awkward transition between existing materials and those used for the improved areas

■ The carriageway is a confusion of different colours, road markings and restricted lane uses, with no physical element or shift in layout to reinforce the separation

Top to bottom:

Green markings are obtrusive and fail to blend in well with the environment

Red markings have been reduced to a single strip but are still effective

The original materials are deteriorating next to the recent improvements

Sutton High Street
Sutton

CASE STUDY

Street type	Suburban main street
Intervention	Environmental improvement and traffic management
Principal authority	London Borough of Sutton

PROJECT RESPONSIBILITIES

Highway authority	LB Sutton
Design	LB Sutton
Project management	LB Sutton
Maintenance	LB Sutton Red Route: LB Sutton for Transport for London

DESCRIPTION

- Busy pedestrian route from train station to shops, bars, offices, etc

- The station is used heavily by commuters and is served by ThamesLink trains running to, and from, central London

- Sutton High Street is the main shopping street for the surrounding area

- The one-way system, which crosses the street, is part of the strategic road network for Greater London and is designated as a 'Red Route'

DESIGN OBJECTIVES

- Improve quality of materials in the street

- Part of the progressive pedestrianisation of Sutton High Street

- Improve the pedestrian environment

- Provide for cyclists and through traffic

Top to bottom:

Part of the street has restricted access for traffic and forms a pedestrian zone during the day

The station provides a key entrance and point of activity on the street

Colour-toned pedestrian crossing facilities have been improved near the station

At Red Route crossings, pedestrians are left isolated in an island while traffic flows freely

VISUAL APPRAISAL

The street has some planting, the more established of which makes an impact on the streetscene. Otherwise it is a hard-edged urban street.

The paving and street furniture are uncoordinated. The whole scheme is confusing visually and has little aesthetic appeal.

At the one-way system crossing points, different paving styles are used and the level of street furniture increases.

The street is well fronted by active properties and has a good sense of enclosure.

QUALITY ASSESSMENT

Comfort & safety	Paving materials provide a smooth comfortable walking surface. One-way system crossings are cluttered and uncomfortable with high traffic speeds.
Domination by functions	Provision has been made for all types of movement. Traffic still dominates in non-pedestrianised areas. Transition into pedestrianised area not handled well.
Visual simplicity	Generally too cluttered — especially at crossings. There is too much street furniture and it is uncoordinated. Paving of every style, colour and bond is present.
Utilities subordinate	Service ducts are visually obtrusive in many areas as the improved paving does not join them — a concrete or tarmac in-fill has been used.
Fitting to character	Pedestrian crossings are not well designed, which detracts from the street's role as a busy pedestrian route.
Ordered for access	Little provision for deliveries or parking in the non-pedestrianised areas. The Red Route classification obviously affects this situation.

KEY ISSUES

- At the junction with the one-way system and Red Route there are poor pedestrian crossings with narrow footways, guardrails and confused routes

- Surface materials of too many colours and styles, resulting in a jumbled mess — no strategy has been developed for the design

- Street furniture is not coordinated in any way and there is too much of it, cluttering the street

- Evening vehicular access to pedestrian zone has led to the zealous use of bollards and signs in a primarily pedestrian area

- Specification and construction standards allowed the scheme to be compromised by poor detailing and construction

Top to bottom:

Paving surfaces disrupted by poor attention to cellar covers — barriers used at red route crossings

Too many paving styles, bonds and colours have been used with no coordination — the pattern here highlights the CCTV camera

Inadequate attention to paving detail at construction stage

Appendix 2.0

Management powers and duties for streets in England

The responsibilities for regulating and maintaining streets forms a complex web of management levels. In practice, the distinctions between authorities' duties are not at all clear cut.

Table A2.1: Powers and duties for motorways and trunk roads in England

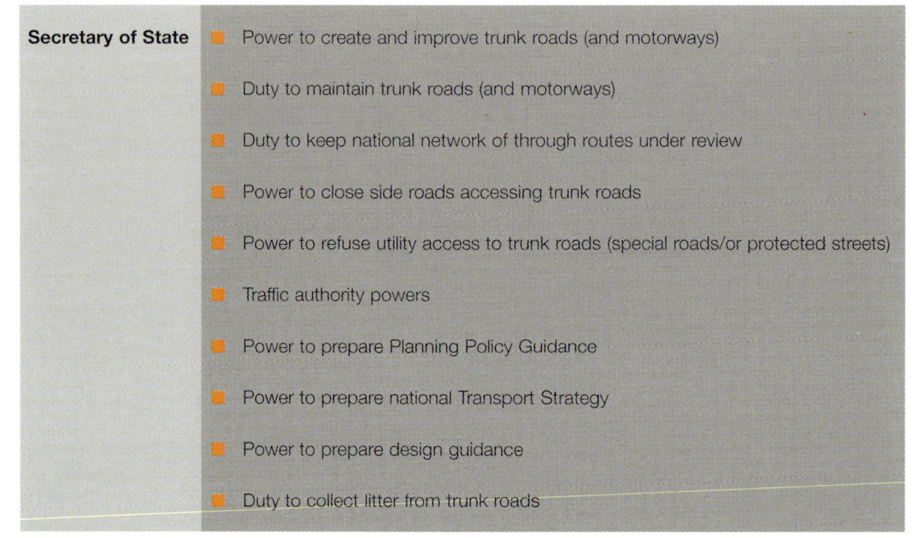

Secretary of State	
	Power to create and improve trunk roads (and motorways)
	Duty to maintain trunk roads (and motorways)
	Duty to keep national network of through routes under review
	Power to close side roads accessing trunk roads
	Power to refuse utility access to trunk roads (special roads/or protected streets)
	Traffic authority powers
	Power to prepare Planning Policy Guidance
	Power to prepare national Transport Strategy
	Power to prepare design guidance
	Duty to collect litter from trunk roads

Table A2.2: Powers and duties for other roads in single tier authorities in England

Unitary Authority	
	Duty to maintain and power to improve all highways (other than motorways and trunk roads, including footpaths and bridleways)
	Traffic authority powers, including powers to restrict and exclude certain types of traffic
	Duty to prepare Development Plan
	Power to prepare Supplementary Planning Guidance
	Duty to prepare Local Transport Plan
	Duty to prepare Community Strategy
	Power to prepare design guidance for road design and adoption
	Powers to adopt new roads as public highways
	Powers to coordinate and inspect works by utility companies
	Power to protect streets from utility access
	Planning authority with control over non-permitted (advertising, telephone kiosks, etc.) development within highways
	Power to erect drinking foundations, horse troughs, seats, litter bins, information signs, kiosks, etc.
	Power to licence markets and traders
	Power to promote or improve the environmental well-being of their area
	Duty to collect litter from all roads (except motorways)

County authorities	• Duty to maintain principal roads and all other local roads (including certain bridges and structures) • Powers to improve highways • Duty to prepare Regional Plan • Duty to prepare Local Transport Plan • Duty to prepare Community Strategy • Power to adopt new roads as public highway • Power to prepare design guidance for road design and adoption • Power to delegate certain powers to District Councils under agency agreements • Traffic authority powers, including powers to restrict and exclude certain types of traffic • Power to protect streets from utility access • Power to promote or improve the environmental well-being of their area
District councils	• Power to maintain footpaths, bridleways and urban 'unclassified' roads (with costs reimbursed from County) • Power to protect streets from utility access • Planning authority with control over non-permitted development within highways (advertising, telephone kiosks, etc.) • Power to erect drinking foundations, bus shelters, horse troughs, seats, litter bins, information signs, kiosks, etc. • Power to licence markets and traders • Duty to prepare Development Plan • Power to prepare Supplementary Planning Guidance • Duty to prepare Community Strategy • Power to promote or improve the environmental well-being of their area • Duty to collect litter from all roads (except motorways)
Parish councils & community councils	• Powers to maintain certain footpaths and bridleways • Powers to erect certain items of street amenities (seats, etc.)

Table A2.3: Powers and duties for other roads in two-tier authorities in England

Table A2.4: Powers and duties for other roads in London authorities

Greater London Authority	
	Duty to maintain and power to improve all metropolitan roads (now Greater London Road Network)
	Traffic authority powers, including powers to restrict and exclude certain types of traffic on all metropolitan roads
	Duty to prepare Transport Strategy
	Power to refuse utility access on protected streets
	Duty to collect litter from metropolitan roads

Boroughs	
	Duty to maintain and power to improve all non-principal roads and other highways
	Duty to prepare Development Plan
	Power to prepare Supplementary Planning Guidance
	Duty to prepare Local (or interim) Transport Plan
	Duty to prepare Community Strategy
	Traffic authority powers, including powers to restrict and exclude certain types of traffic on non-metropolitan roads
	Planning authority with control over non-permitted development within highways (advertising, telephone kiosks, etc.)
	Power to adopt new roads as public highway
	Power to prepare design guidance for road design and adoption
	Powers to coordinate and inspect works by utility companies
	Power to erect drinking foundations, horse troughs, seats, litter bins, information signs, kiosks, etc.
	Power to licence markets and traders
	Power to promote or improve the environmental well-being of their area
	Duty to collect litter from all borough roads

Appendix 3.0

Lessons from abroad

Marseille France

A modern public realm within an historic setting — Cours Estienne d'Orves

Marseille is the third largest city in France with a population of 800,000. Located between the Mediterranean Sea and the southern Alps, the city continues to be France's principal port. Since the mid-1990s, investment has focused on improving public spaces and reducing the impact of vehicles in the centre.

Street ownership and management

Bollards and gates, separate pedestrians from traffic, enabling a level surface and a greater feeling of pedestrian importance

The public realm in France is administrated by the local 'commune', which is the lowest government tier in France. State roads are classified as '*Nationale*,' and others as '*Départementale*,' which belongs to the *Département*. Most roads in, or passing through, a town are owned and managed by the commune. Land in France is *cadastre*, i.e. registered and numbered according to plots. As such, the public realm is land that is not numbered and is public or *non cadastré* by default.

All streets and street furniture (bollards, traffic lights, signs, etc.), belong to the commune. However, each local authority department, or commune, has responsibility for its own street furniture, though all departments are meant to coordinate their works.

One-way system with parking either side of carriageway — Bouelvard Longchamp

Since 1995, a youth employment scheme has provided for uniformed street wardens, known as *Mediators*, to look after the public realm in cities across France. Employed by the public sector, they are equally funded by the state and the city. Their benefits are two-fold, in assisting people on the street and in reporting problems with street amenities back to the relevant authorities.

Legislative framework

The role of the local authority, with regard to the management of the public realm, is written in the Code des Communes. Marseille has its own city highway rules (*Réglement de voiries de la Ville de Marseille*) based on national regulations. Added to this, the *Code de l'Urbanisme* (Town Planning Code), although not definitive, may also outline certain aspects relating to streetscape, although it is primarily concerned with the regulations and rights to build, such as zoning, building lines, etc. The regulatory framework is not detailed but is very clear regarding management roles, leaving all street issues in the hands of the commune.

The state has also asked communes to prepare Urban Movement Plans (*Plans de Déplacements Urbains*), which outline a framework of all modes of transport and where priorities lie within their respective developments. The *Préfet*, who represents the state interest at the local level, has a duty to ensure that this is carried out. The Plan gives the spirit of what is to be achieved and favouring one mode of transport's development (e.g. pedestrian) can be given weight by the Plan.

Implementing street improvements

Until recently, urban design strategies were produced by local area (*quartier*). For example, in the Panier area, the strategy helped to ensure the refurbishment of building façades and paving. Particular sections of the area were classified as 'standard' or 'sensitive'. This has enabled separate areas to be treated differently according to their particular design needs, while taking an overall area approach.

Only 15 years ago, the city set up the Urban Development Department. This is charged with the responsibility for large infrastructure projects, and has created a public realm masterplan that draws together earlier area-wide projects and key local projects. This document is due to be approved by the Council and will be published. It outlines the key issues of good urban analysis required at conception, of competent construction, and of sustainably financed management and maintenance.

The Urban Development Department has also begun to compile a Public Realm Strategy (*Schéma d'espace public*). This is a scheme intended to gradually improve the streetscape in Marseille. It coordinates all street furniture by specifying a suite of agreed products based around a core set of bespoke designed furniture that is unique to the city. However, the Traffic Department and Urban Development Department often have conflicting views on street design.

Top to bottom:

Pedestrian/car segregation through the use of low-rise bollards — Place aux huiles

The old town with narrow streets — Quartier du Panier, Rue du Refuge

Conflict of uses in a primarily pedestrian environment — Bir-Hakeim

Street improvement projects largely emanate from commune politicians, although are often first proposed by technical officers. These are then taken forward by the department teams, who define the programme along with the elected representative. All departments must be consulted on each project, which will involve the traffic, green spaces, cleaning, location (for kiosks, etc.) and lighting services, as well as the fire department, though not the police department.

While the Town Hall will finance improvements as part of its mandate, poor areas may receive subsidies. Under the *Politique de la Ville* (City Improvement) scheme, the state will fund public realm improvements. The European Union, under Regional Development Fund Objective 2, has also allowed funds to be invested in Marseille, classified as a previous industrial region in current conversion.

Left to right:

Segregated bus lane with central island parking bays — Rue de Rome

Vehicular access route to the front of buildings on a square

The main pedestrianised street in the town centre — Rue Saint Ferréol

Top Left:

Imposing street furniture appropriate to wide French avanues (hanging wires are for trolley-bus lines) — Square Léon Blum

Top to bottom:

Temporary bollards and traffic management prior to redevelopment and new library — Cours Belsunce

Marseille's hilly topography creates difficult road junctions and highway configurations

Planning and supervising utilities

Anyone wishing to intervene in the public realm has to apply for permission from the commune, they then pay rent during the period of the intervention. A specific department in the Town Hall deals with all applications, the *Service des emplacements* (Location Service). This department must consult all other departments within the Town Hall when an application is received. This system applies to both above ground (scaffolds, market stalls or temporary structures) and underground works. Statutory undertakers therefore have to ask permission before effectively renting underground space.

Like the UK, privatisation of utility companies has affected the system for the granting of permissions, as the commune has seen a vast increase over recent years in both the numbers of permission requests, and the range of companies applying. At times, the service is overwhelmed and permissions are granted without due consideration. This problem is set to increase, while at the same time there is now more recognition of the need to take more care of the public realm.

All utilities networks and street furniture are registered on a database, although the administration is so large that information does not always circulate properly nor does it always get updated regularly, with the concomitant problems that this brings. Poor remedial work is ongoing and the department in charge of looking after this work lacks staff and investment for this, although the Mediators (street wardens) do assist with the reporting of defects.

Copenhagen Denmark

Copenhagen Commune has a population of 500,000 and, despite the fact that it is the capital city, still retains a low-rise character. Located on the Øresund, which separates it from Sweden, water is a prominent feature throughout. The 1990s were a period of restoration, characterised by a considerable urban renewal programme of public spaces, for which the city can now be seen as a model.

Street ownership and management

Responsibility for public roads in Denmark is shared three ways between state, county and municipalities. Since 1972, the Roads Directorate has been responsible for the Danish National Road Network. It is responsible for the provision, operation and maintenance of the main road infrastructure of motorways, other national roads, and their bridges and tunnels. Its remit extends from bypasses, environmentally prioritised alignments, bicycle paths, pavements and snow clearing. It also carries out research and development of new materials, management systems for road and bridge maintenance, road safety, and environment and traffic informatics. The Directorate thus draws up national standards to be adopted by local autrhorities.

Counties and municipalities are each responsible for maintaining, improving, and removing accident spots and cleaning off snow on their respective stretches of road. They also lead local campaigns to encourage more responsible traffic behaviour. Details extend to ensuring that the grass verges along the roadsides are cut, and lay-bys kept clean and tidy, as well as snow clearing and gritting. Physical planning in counties and municipalities is coordinated by the County in order to make measures appropriate and avoid environmental damage.

At the local level, the Copenhagen Municipality handles planning, orders and organises invitations for tenders for cleaning, and maintains its own roads and cycle paths, and keeps road surfaces clear. The Public Roads & Park Agency, *Vej & Park*, within the Copenhagen Municipality, is responsible for traffic planning and management, including signals, traffic census and traffic safety.

Landowners are responsible for cleaning, snow clearing and salting or gravelling pavements. By inspection and orders, the municipality ensures that private communal roads are in a condition that the size and nature of the traffic require.

Left to right:

A modern façade in an historic setting — the Danish Centre for Architecture on Strandgarde

Knud Holsher's bus shelter, specially designed for the Metropolitam Bus Company, HT — HC Andersens Boulevard

Legislative framework

Local plans are implemented and drawn up on the initiative of the Municipality, possibly at the request of landowners, within the framework for the use of the areas. They contain guidelines, among other things, as to how the open spaces of the area are to be located and laid out.

Implementing street improvements

Since 1962, when Copenhagen's main street, Strøget, was pedestrianised, the Municipality of Copenhagen has been actively pursuing the gradual reduction of parking spaces in favour of improving the public realm for the enjoyment of pedestrians. The Municipality maintains a policy of not increasing traffic in the town centre, and this has been successfully achieved so that the current volume is at the same level as that from 1970. Parking provision in the city centre has, in this way, been reduced by 2–3% annually. This has been undertaken so gradually that the population has gladly accepted it, as they have discovered new places of quality. Where once all public squares were merely car parks, they stand today as proper city squares in which real investment has been made on providing a quality streetscape with quality street furniture and materials.

It therefore does not seem to be a political issue, as the slow pace in reduction has meant that people have also gradually left the car for other modes of transport. This has apparently also been aided by the enthusiasm, as much from the side of the municipal architects as from the engineers, for the need to consider people and buildings of quality as a first measure in all development. Copenhagen's slow change from a car-dominated city to a pedestrian-friendly environment has even allowed the totally unexpected café terrace culture to reach Scandinavia.

Copenhagen has a very firm policy on public space and on traffic. Motorists are compelled to obey the rules, thereby ensuring that there is no illegal parking or illegal driving practices, making the city that much safer and environmentally pleasing.

Planning and supervising utilities

A permit must be applied for to use both public and private communal roads for various activities, for example containers or services. The permit is issued by the Public Roads and Park Agency at the Copenhagen Municipality.

A feature of Copenhagen is the availability of free city bikes (*Bycyclen*), introduced to Copenhagen by the City, which invested heavily in providing cycle paths alongside many of the main streets. A number of sponsors fund the scheme, which demands City Bike Patrollers, who maintain the bikes and bike stands every day, making on-the-spot repairs when necessary.

Top to bottom:

City Bike stand — Laksegarde

Arne Jacobsen's Danish National Bank — Niels Juels Gade

Street clutter avoided by suspended street and traffic lights — Holmens Kanal

Freiburg Germany

Freiburg im Breisgau is a university city in south-west Germany, close to the Swiss and French borders. It has a population of 135,000, with a further 60,000 living in adjacent suburbs and villages. It is well known (at least within Germany) because in recent years the Green Party has held the balance of power on the local council, with 20–25% of seats, and has exercised that power to promote policies encouraging sustainability in development and transport.

Street ownership and management

Roads and streets in Germany are managed according to a three-tier hierarchy:

- motorways — Federal Government
- main roads and county roads — the 16 Lander (Freiburg is in the Lander of Baden — Württenberg)
- city streets — the local authority (Grossestädte) where the town or city is 80,000 or more.

Freiburg has responsibility for all the roads and streets within its boundaries, not unlike the agency agreements between County Highway Authorities and District Councils in England. However, it would appear that it exercises that responsibility with a greater degree of independence than the English system allows.

The division of responsibility for streets and roads in cities of over 80,000 population

Legislative framework

Design codes for road and street layouts are published by the Federal Government, in particular *Rechtlinien Für die Anlage von Strasse RAS*, which is the German equivalent of *Design Bulletin 32*. The Lander also publish design guides.

An interesting aspect of German practice is that Local Communities can pay to have variations on the standard road or street treatment. This appears to apply more to rural communities than to local sectors of a city, such as Freiburg.

Implementing street improvements

Cities such as Freiburg each have a Traffic Development Plan, published following extensive consultation. Given Freiburg's green credentials it is hardly surprising that its plan strongly promotes sustainable means of transport — walking, cycling and public transport (especially trams). Over the years there has been a slight reduction in car usage, and a significant increase in travel by bus, tram and bicycle. The figures suggest that the main change has not been people forsaking their cars for cycling and public transport, but a switch from walking to those forms of travel. But the key thing, in the eyes of the local authority, is that the number of car journeys has been held in check.

Journey types in Freiburg	1982	1999
by foot	35%	22%
by bicycle	15%	26%
by bus/tram	11%	18%
by car	38%	32%

Traffic calming projects, pedestrianisation and public transport improvements are all the responsibility of the local authority, as regards both strategic planning and detailed implementation. Improvements have ranged from the repairing of inner city streets to a major scheme linking the city centre to the north-west suburbs by means of a new tram route. This new route connects directly to the main railway station, where there is also a purpose built cycle hire facility. These projects, aimed at curbing car use in the city, are a clear result of local political leadership. They are somewhat at odds with the policies still being followed in the surrounding Lander, where it is assumed that a gradual increase in car use over the next 10–15 years should be provided for.

Annual expenditure in Freiburg on streets and roads can be summarised under these headings:

- new works, including pedestrianisation, park and ride facilities and street improvements — 10 million euros per annum
- street maintenance — 3 million euros per annum
- maintenance of tramway infrastructure — around 5 million euros per annum

Planning and supervising utilities

A notable aspect of streetscape management in Freiburg is the close supervision exercised over utility companies. Essentially, this takes three forms:

- regular planning meetings to discuss utility renewals
- a fee paid by the utility company to the city for opening up works
- a bond of 2% of the contract price is retained by the city to fund repairs if reinstatement works are inadequately completed.

This tighter control mechanism could provide lessons for UK practice in curbing the excesses of utility companies' interventions.

Top to bottom:

Interchange between train, bus and tram at the main station

A typical street improvement — Rotlaubstrasse

Tram route on Kaiser-Joseph Strasse

Appendix 4.0

Legislation and guidance affecting the street

This summary shows the key statutes, regulations and design guides of the past and present that influence the current spatial arrangement or management of streets. It is intended as a reference, but is not to be viewed as comprehensive.

Marking and furniture

Statutes

Public Health Acts (Amendment) Act	1890
Public Health Act	1925
Local Government (Miscellaneous Provisions) Act	1953
Planning (Listed Buildings and Conservation Areas) Act	1990

Regulations and orders

Town and Country Planning (Control of Advertisements) Regulations	1992
Traffic Signs Regulations and General Directions SI 1994 No. 1519	1994
General Permitted Development Orders	1995

Non-statutory design guides

Government		
Traffic Signs Manual		1967–1997
PPG19: Outdoor Advertisement Control		1992
Local Authority		
Local Authority Design Guides		Various
Government agencies		
Urban Design Compendium	EP	2000
Other		
BS5489: Road Lighting	BS	1992
BS873: Road Signs	BS	
BS EN 1790: Road Markings	BS EN	2000
BS7263 Parts 1 and 3: Paving	BS	2001
BS7533 Parts 4 and 6: Pavements of clay, stone and concrete	BS	1998, 2000
BS EN 1343: Kerbs of natural stone	BS EN	2000

Traffic

Statutes

Town and Country Planning Act	1990
Highways Act	1980
Road Traffic Regulation Act	1984
New Roads and Streetworks Act	1991
Transport and Works Act	1992
Road Traffic Act	1988
Traffic Calming Act	1992
Greater London Authority Act	1999
Town Police Clauses Act	1847
Metropolitan Public Carriage Act	1869

Regulations and orders

Pelican and Puffin Pedestrian Crossings and General (Amendment) Directions SI 1998 No. 901	1998
Zebra Pelican and Puffin Pedestrian Crossings Regulations and General Directions SI 1997 No. 2400	1997
Highway (Road Humps) Regulations SI 1996 No. 1483	1996
Traffic Signs (Temporary Obstructions) Regulations	1985
Pelican Pedestrian Crossing Regulations and General Directions	1987
Highways (Traffic Calming) Regulations	1993
New Street Byelaws (Extension of Operation) Order	1997

Non-statutory design guides

Government

PPG6: Town Centres and Retail Developments	DETR	1996
Going to town: Improving town centre access — companion Guide to PPG 6	DTLR	2000
PPG13: Transport	DETR	2001
Circular Roads	DoT/HA	1970–1996
Traffic Advisory Leaflets	DETR/DoT	1989– 2001
Local Transport Notes	DETR	1986–1998
DB32 Residential Roads and Footpaths	DoE	(1977) revised 1992
Places, Streets and Movement — Companion Guide to DB32	DETR	1998
Design Manual for Roads and Bridges	DoT/HA	1994

Local Authority

London Cycle Network	LB Kingston-upon-Thames, RB Kensington and Chelsea *et al.*	2000
Local Authority Design Guides		Various

Government Agency

Urban Design Compendium	EP	2000

Other

Guidelines for providing for journeys on foot	IHT	2000
Guidelines for Safety Audit of Highways	IHT	1990
Designing for Deliveries Freight	TA	1983
Transport in the Urban Environment	IHT	1997
The Design of Roundabouts	IHT	1995
Transport in the Urban Environment	IHT	1997
Cycle-friendly infrastructure	IHT	1996
National Cycle Network — Guidelines and Details	Sustrans	1997

Services

Statutes

Light Railways Act	1912
Harbours Act	1964
Post Office Act	1969
British Telecommunications Act	1981
Cycle Tracks Act	1984
Gas Act	1986
Electricity Act	1989
Water Act	1989
British Technology Group Act	1991
Water Industry Act	1991
New Roads and Street Works Act	1991
Transport and Works Act	1992

Regulations and orders

The Street Works (Qualifications of Supervisors and Operatives) Regulations	1992
The Street Works (Inspection Fees) Regulations	1992
The Street Works (Inspection Fees) (Amendment) Regulations	1998
(revoked and replaced *in England only* by 2001 Regulations — see below)	
The Street Works (Reinstatement) Regulations	1992
available with: The Street Works (Reinstatement) (Amendment) Regulations	1992

The Street Works (Sharing of Costs of Works) Regulations	1992
(revoked and replaced *in England only* by 2000 Regulations — see below)	
The Street Works (Maintenance) Regulations	1992
The Street Works (Notices) Order	1992
The Street Works (Registers, Notices, Directions and Designations) Regulations	1992
The Contracting Out (Highway Functions) Order	1995
The Street Works (Registers, Notices, Directions and Designations) (Amendment) Regulations	1995
The Street Works (Registers, Notices, Directions and Designations) (Amendment No. 2) Regulations	1995
(revoked and replaced by Amendment No. 3 Regulations — see below)	
The Street Works (Registers, Notices, Directions and Designations) (Amendment No. 3) Regulations	1995
The Local Authorities Traffic Orders (Procedure) Regulations	1996
The Street Works Register (Registration Fees) Regulations	1999
The Street Works (Registers, Notices, Directions and Designations) (Amendment) Regulations	1999
The Local Authorities (Contracting Out of Highway Functions) Order	1999
The Street Works (Sharing of Costs of Works) (England) Regulations	2000
The Local Authorities (Contracting Out of Highway Functions) (England) Order	2001
The Street Works (Inspection Fees) (Amendment) (England) Regulations	2001
The Street Works (Charges for Unreasonably Prolonged Occupation of the Highway) (England) Regulations	2001

Codes of practice (street works)

Specification for the Reinstatement of Openings in Highways	1992
Measures Necessary where Apparatus is Affected by Major Works (Diversionary Works)	1992
Code of Practice for Inspections	1992
Safety at Street Works and Road Works — A Code of Practice	2001
Code of Practice for the Coordination of Street Works and Works for Road Purposes and Related Matters First edition	1992
(from 1 April 2001 applies in Scotland and Wales only)	
Code of Practice for the Coordination of Street Works and Works for Road Purposes and Related Matters: APPENDIX E	
(From 1 April 2001 applies *in Wales only*; in England from that date it was superseded by the new version of Appendix E incorporated in the 2nd edition of the Code — see the next item)	
Code of Practice for the Coordination of Street Works and Works for Road Purposes and Related Matters	Second edition 2001
(From 1 April 2001 applies *in England only*, replacing the 1992 edition)	

Non-statutory design guides

Government

Planning for Telecommunications	DETR	1999
PPG8: Telecommunications	HMSO	1992

Government Agencies

Urban Design Compendium	EP	2000

Local authority

Local Authority Design Guides		Various

Other

Soakaway Design	BRE	1991
Drainage capacity of BS Gullies	TRL	1984
Hydraulic efficiency and spacing of BS gullies	TRL	1969
Sewers for Adoption	WSA	1995
Recommended positioning of utilities apparatus	NJUG	1997
Guidelines for utility services close to trees	NJUG	1995

Public health and environment

Statutes

Public Health Act	1925
Highways (Miscellaneous Provisions) Act 1961	
Licensing Act	1964
Civic Amenities Act	1967
Control of Pollution Act	1974
Food and Environmental Protection Act	1985
Local Government, Planning and Land Act	1980
Disabled Persons Act	1981
Local Government (Miscellaneous Provisions) Act	1976, 1982
Building Act	1984
Removal and Disposal of Vehicles Regulations	1986
Town and Country Planning Act	1990
Environmental Protection Act	1990
Planning (Hazardous Substances) Act	1990
Planning (Consequential Provisions) Act	1990
Planning and Compensation Act	1991
Disability Discrimination Act	1995
London Local Authorities Act	1995
Local Government Act	2000

Regulations and orders

Town and Country Planning (Tree Preservation Order) (Amendment) and (Trees in Conservation Areas) (Excepted Cases) Regulations		1975
Building Regulations (Parts B5, F, M)		1995–2000
Town And Country Planning (Use Classes) Order		1987

Non-statutory design guides

Government

PPG 1: General Policies and Principles		1997
PPG 10: Planning and Waste Management		1999
PPG 24: Planning and Noise		1994
'By Design' Urban design in the planning system: towards better practice	DETR/CABE	2000

Other

Access for Disabled People: Design Guidance	CoEf Handicapped 1985	

Public order

Statutes

Public Order Act	1986
Local Government (Miscellaneous Provisions) Act	1976
Criminal Justice Act	1981
Criminal Justice and Public Order Act	1994
Crime and Disorder Act	1998

Non-statutory design guides

Planning Out Crime (Circ 5/94)	DoE	1994
Public Rights of Way	DoE	1993
Urban Design Compendium	EP	2000
Secured By Design	Association of Chief Police Officers	1994

Specifications

Design Manual for Roads & Bridges	DoT/HA	1994
Volume 6 Road Geometry:		
TD 09 Highway Link Design		1993
TD 27 Cross Sections and Headroom		1996
TA 85 Guidance on minor improvements		2001
TD 22 Layout of grade separated junctions		1992
TD 48 Layout of grade separated junctions		1992

TD 16 Geometric design of roundabouts	1993
TD 78 Design of road markings	1997
TD 50 Layout of signal controlled junctions	1999
TD 39 Design of major interchanges	1994
TD 40 Compact grade separated junctions	1994
TD 42 Major/minor priority junctions	1995
TD 41 Access to all-purpose trunk roads	1995
TA 23 Size of roundabouts and junctions	1981
TD 36 Subways for pedestrians and cyclists	1993
TA 66 Police observation platforms	1995
TA 69 Location and layout of lay-bys	1996
TA 57 Roadside features	1987
TA 81 Coloured surfacing of road layouts	1999

Appendix 5.0

Potential changes to guidance

Table A5.1: Design Manual for Roads and Bridges — key documents

This appendix lists examples of current guidance, notes their defects in terms of treating streets for multiple uses and suggests ways of rectifying this imbalance.

Document	Specific failure in guidance	Proposed solutions
Design Manual for Roads and Bridges TD 42/95 Part 6 *Geometric design of major/minor priority junctions*	Assumption that the movement of traffic is the main concern in junction design 'advantage…is that through traffic on the major road is not delayed'. No mention of advantages for pedestrians or cyclists.	Revise with a more balanced view highlighting advantages for all road users in urban areas where pedestrian and cycle needs are higher.
Design Manual for Roads and Bridges TD 16/93 *Geometric design of roundabouts*	Main objective to provide safe traffic interchange with minimum delay is not appropriate to all urban areas. Mini roundabout design recommended for urban situations without mention of pedestrians who find them difficult to cross. The inherent conflict with pedestrians created by roundabouts is hugely understated. Where it is mentioned, pedestrian facilities are promoted away from the junction point, and guardrails are required to prevent 'indiscriminate' crossing. Suggests that smaller roundabouts can be attractive in sensitive areas without further amplification on this important issue.	More recognition required for the difference between urban and rural roundabout design, in the context of needing to prioritise pedestrian movement more highly in urban situations. Multi-lane pedestrian crossings should be encouraged without major recourse to guardrailing. Mini and small roundabout signage can be as intrusive as the roundabout itself. Traditional townscapes do not accommodate roundabouts comfortably.
Design Manual for Roads and Bridges TD 50/99 *Geometric layout of signal-controlled junctions and signalised roundabouts*	While pedestrian movements are considered in the design guidance, the proposals have a vehicle flow bias, with recommendation that pedestrian crossings are located beyond the limits of the junction radius Crossing designs do not distinguish between a new 'greenfield' build design and a design fitted into existing urban areas, so that where space allows, large junctions with many islands are encouraged.	Emphasise the need of contribution to urban quality of pedestrian comfort at crossings. High pedestrian flows should allow in-line crossings, fewer islands and tighter corner radii. Clearly distinguish between urban trunk and local road junction design. Also remove the premise that availability of space should be utilised. If the junction can be made to work in a tightly restricted urban situation, then it should be able to be replicated in a new build situation at the same size.

Table A5.2: Design Bulletin 32 — key issues

Document	Quality streetscape issues	Possible action
Design Bulletin 32	■ Definitions of streets classified as 'Primary Distributors', etc., talking only of the movement of traffic and number of homes served. No mention is made of other users of streets and how they fit into the hierarchy. ■ No consideration of 20 mph zones is allowed for on other than minor access roads. Inconvenience to motorist of speed restaint measures is often emphasised. Phraseology based on driver acceptance. ■ Assumes shared surfaces only work for up to 50 dwellings. ■ Due to the nature of the document and how it was intended to fit into highway design structure, the guidance given only applies to streets in single use residential areas. It is used in most situations, however, due to a lack of other similar guidance for different situations.	■ Revise definition of streets to take into account their more shared-use nature. This is particularly true when talking about primarily residential areas, as DB32 does. ■ Allow for 20 mph zones in all areas. Emphasis on convenience and care of motorist when considering speed restraint can be reduced due to increased societal exposure to traffic calming. ■ Align with home zone advice, which allows for 10 mph to serve 125 dwellings (worked on 100 vph at peak times) ■ Withdraw DB32 and produce new document on the design of 'urban streets', or similar, which provides guidance for design in differing land use contexts from urban core to suburban edge, taking into account all users of the street.

Table A5.3: Local Transport Notes — key documents

Document	Quality streetscape issues	Possible action
Local Transport Notes 1/95 * *Assessment of pedestrian crossings*	■ Text talks of crossings as amenities. ■ 'The manual method for estimating the difficulty of crossing relies on judgement by an experienced traffic engineer'.	■ Rewrite with a more sympathetic tone, crossings are necessities for many street users. ■ Suggest pedestrian audits by varying street users to assess the level of crossing difficulty, as it will vary considerably for a pensioner or small child from that of a traffic engineer.
Local Transport Notes 2/95* *The design of pedestrian crossings*	■ Text has an underlying assumption that crossings should be made so as not to inconvenience the driver. Use of non-staggered refuges are not recommended, for example as they may confuse drivers — possibly more likely that they allow traffic to move more freely at the expense of pedestrian comfort	■ Revise, with emphasis on improving pedestrian traffic flow and comfort. Facilities should be designed to best suit pedestrians, concerns over traffic should be secondary

* Also TA 68/96 in Design Manual for Roads and Bridges

Table A5.4: Traffic Advisory Leaflets — key documents

Document	Quality streetscape issues	Possible action
Traffic Advisory Leaflets	■ Documents generally are of poor graphic and photographic quality, sometimes highlighting average or low build quality schemes as exemplars. This reflects the apparent design value and consideration given by central government. Only best practice in engineering and streetscape terms should be illustrated, and there are plenty of examples today. ■ Unclear as to the weight of the document, are they guidance or to be followed to the letter (authorities want more prescriptive guidance and will follow documents to the letter even if that is not their intended function). ■ The documents are generally poor in their layout and graphic style (drop shadow boxes on images, for example) resulting in an unprofessional appearance. ■ Case study leaflets tend to be clearer than others. They are more focused in their attention and the images used are less open to interpretation.	■ Need to provide higher grade of design as exemplary, so that if budgets are restricted, the principles can be followed, but the starting point example is of very high quality. ■ Vet documents by an experienced landscape/streetscape advisor prior to publication. ■ Use sketches if quality photographs of real schemes are unavailable. ■ Clarify statutory or advisory weighting and position of documents within the text. Explain the use of urban design/building alignment/pedestrian desire lines/other street furniture, public art, etc., as tools to enhance the streetscape as well as carry out traffic management. ■ Simplify layout and improve graphic style — later TAL are better but more improvement is necessary. ■ Build TAL case study coverage of good practice.
Traffic Advisory Leaflet 1/87 *Measures to control traffic for the benefit of residents, pedestrians and cyclists*	■ A set of crudely sketched area-based traffic calming possibilities with great intent. Emphasis is on low cost, though the intention is to provide more pleasant conditions.	■ Revise with better illustrations, less emphasis on cost, more on value.
Traffic Advisory Leaflet 7/93 *Traffic calming regulations*	■ The top level document to a number of traffic calming and speed restraint design guides emanating from the Traffic Calming Act 1992, allowing promotion of safety and improvement of the environment. The document describes, with illustrations, the key traffic calming elements available to designers. ■ Pinch points, chicanes, islands, build outs and speed cushions are illustrated in inappropriate materials, and poorly detailed in relationship to existing townscape and materials. This gives the idea that they are acceptable for others to use. ■ The premise is not clear that these are retro-fitting measures. Much new road building fails to follow *Places Streets and Movement* and includes these speed restraint measures when speed should be controlled by block layout design, building alignment and positioning of junctions.	■ Illustrate with more environmentally sensitive schemes. ■ Clarify the design intention of new streets should follow that set out in *Places Streets and Movement*.

Document	Quality streetscape issues	Possible action
Traffic Advisory Leaflet 12/93 *Overrun areas*	■ Corner radius over runs are difficult for pedestrians. *Places Streets and Movement* demonstrates how tighter radii may be accommodated in some streets for new designs, and on some lightly trafficked streets it may be acceptable to allow larger vehicles to cross the centreline of the side road. ■ Images shown have detailed design failings, such as vehicle tracks on verge, a roundabout whose edge has failed, and a light duty telecom inspection chamber cover in the over run	■ Draw attention to other design potentials before advocating over runs. ■ Revise with well-detailed designs illustrated, which are substantially more sympathetic to their townscape.
Traffic Advisory Leaflet 13/93 *Gateways*	■ Schemes shown are heavy handed and images are of poor quality. All the schemes shown urbanise otherwise rural locations with fittings that are usually associated with urban areas.	■ Revise with well-detailed designs sympathetic to townscape.
Traffic Advisory Leaflet 4/94 *Speed cushions*	■ Schemes shown are unsympathetic to their surroundings and offer little benefit to pedestrians.	■ Revise with better examples and more emphasis on suitability of this kind of traffic calming element.
Traffic Advisory Leaflet 9/94 *Horizontal deflections*	■ A major speed restraint treatment illustrated with photographs of experimental layouts with plastic dividers and incredibly brutal and unsophisticated installations in streets. Poor signage, materials selection (kerb and paving treatments in non-matching materials, timber planters and posts in inner urban locations), all unsuitable to the townscape settings. ■ Cycle lanes shown inside deflections are difficult to maintain. ■ Steps to integrate the revised drainage patterns caused by build outs should be dealt with, not avoided, by leaving channels open or covered in the safety channel.	■ Revise illustrations, provide more varied use of deflections to also assist pedestrian crossings on desire lines. ■ Mention the need to align deflections with townscape and building alignment events to reinforce and integrate into the street character, rather than to appear as an add-on. ■ Deflections should use matching materials to adjoining areas.
Traffic Advisory Leaflet 7/95 *Traffic islands for speed control*	■ The guide points to the use of islands without pedestrian facilities for speed reduction to reduce accidents and improve the local environment. The latter can only be improved if the design is integrated in materials and form with the surrounding streetscape and built form. The measures illustrated severely impact on the streetscape with poor quality finishes, materials that do not match and that are not maintained.	■ Mention the need to align deflections with townscape and building alignment events to reinforce and integrate into the street character, rather than to appear as an add-on. ■ Revise with well-detailed designs illustrated, which are substantially more sympathetic to their townscape.

Table A5.4: Traffic Advisory Leaflets — key documents (continued)

Document	Quality streetscape issues	Possible action
Traffic Advisory Leaflet 12/97 *Chicane Schemes*	Many of the solutions shown are poorly detailed, poorly integrated into existing streetscape (especially poor matching of materials), unsympathetic to townscape and without means of access for sweeping which are potentially dangerous to cyclists.	Revise with sympathetic well-detailed designs (see also TAL 9/94 actions above).
Traffic Advisory Leaflet 9/99 *20 mph speed limits and zones*	Speed controlling measures are limited to the use of humps, raised junctions, cushions, horizontal deflections in the highway (as TAL 7/93, 9/94 and 12/97), and bends. The effect of building lines and pinch points created by these is not explored. The proposals suggested are largely retro-fit solutions for existing situations	The document is very much a defensive measure against vehicle speed. Create better clarity about the relationship of urban form, activity and buidling line to highway speed design. Describe more how this can be integrated as suggested in *Places, Streets and Movement*.

Table A5.5: Transport for London Design Guidance — key documents

Document	Quality streetscape issues	Possible action
Walking Schemes: *Good Practice Guidance for London*	Schemes shown are heavy handed, with isolated pedestrian islands, guardrailing and uncoordinated signage and street furniture.	Revise with illustration of more townscape and pedestrian-comfort sensitive design examples.

Table A5.6: *Highway Code* — key issues

Document	Quality streetscape issues	Possible action
Highway Code	'In urban areas there is a risk of pedestrians…stepping unexpectedly into the road.' The guidance, while apparently more biased to pedestrians on the front cover, lists driver behaviour towards pedestrians in only six paragraphs about vulnerable road users. The road is very much portrayed as for the motorist.	Revise with more emphasis on the shared nature of streets in urban areas. Distinguish between roads in urban and rural areas more strongly. Provide more emphasis on driver behaviour towards pedestrians expected at side road junctions, raised speed tables with crossings (courtesy crossings), and crossovers.

Appendix 6.0

Consultees, project steering group and acknowledgements

Consultees

ORGANISATION

AA	Paul Watters
Association of British Insurers	Mary Francis, Director General
Association of Chief Police Officers	Miss Marcia Barton, OBE
Association of Town Centre Management	Alan Tallentire
Association of Town Centre Management	Elaine Robertson
Birmingham City Council	Phil Crabtree
British Property Federation	William McKee
British Road Federation	
Civic Trust	Ben Webster Liz Wrigley
Cotswold District Council	Kim Cooper, Director of Planning
Countryside Agency	Fiona Fraser-Boulton
Countryside Agency	Denis Seabeck
CPRE	Donald Mitchell
CPRE	Richard Bourne
Confederation of Passenger Transport	Steve Salmon
CZWG Architects	Piers Gough
Derbyshire County Council	Barry Joyce
Dorset County Council	Ian Madgwick
Epsom and Ewell Borough Council	Eileen Thomas
Freight and Transport Association	Geoff Dossetter
Great London Authority	Mark Brearley
Highways Agency	Mima Garland Mem Baybars
Highways Authority Utilities Committee	Terry Hecquet
House Builders Federation	Andrew Whitaker Mike Newton
HTA Architects	Amanda Taylor
Institute of Historic Buildings Conservation	Eddie Booth
Institution of Highways and Transportation	Carlton Robert-James

Incorporated Society of British Advertisers	Bob Wootton
Kent County Council	Theresa Trussell
Lichfield District Council	John Colburn
London Borough of Camden	Robin Harper
London Borough of Tower Hamlets	Ludo Reid
London Borough of Tower Hamlets	Tom McCourt
London Borough of Tower Hamlets	Joe Tavernier
Metropolitan Police	Inspector Brian Howat Calvin Beckford, Crime Prevention Design Officer (Camden)
Middlesbrough Council	Phil Harper, Director of Planning
National Joint Utilities Group	Irene Elsom
Northampton Borough Council	John Harvey Jurek Waluda
Outdoor Advertising Association	Matthew Carrington
RAC	Matt Flutes Manager of Travel Information
Road Users' Committee	Joyce Wethered
RoSPA	John Howard
Royal Town Planning Institute	Margaret Catran
Sheffield City Council	John Charlton, Head of Street Force David Young Richard Watts
Shropshire County Council	Rob Surl
South Shropshire District Council	James Caird
St Edmundsbury Borough Council	Ian Poole
Sustrans	Katie Dixon
Swindon Borough Council	Stephen Hardy
Transport for London	Nicola Cheetham Mem Baybars
Transport 2000	Lynn Sloman
Urban Initiatives	Paul Murrain
Urban Design Alliance	Robert Huxford (ICE)
Westmister City Council	Rosemary McQueen
Wilcon Homes Ltd	John Weir

Steering group

Organisation	Name
CABE	Dickon Robinson
CABE	Robert Bargery
CABE	Ben Castell
DETR	Peter Ellis
DETR	Mark Plummer
DETR	Peter Matthew
DETR/Disabled Persons Transport Advisory Committee	Tim Pope
English Heritage	Charles Wagner
Living Streets (formerly Pedestrian Association)	Ben Plowden
Disabled Persons Transport Advisory Committee/NorthLanarkshire Council	Graham Lawson
Local Government Association/Dudley Metropolitan Borough Council	Paul Watson
Urban Design Alliance	Lynda Addison

Additional acknowledgements

Atelier de l'Espace Public, Ville de Marseille, France	André Rulliere
AFA JC Decaux, Denmark	Kim Axelsen Ken Warpole
Birmingham City Council	Phil Crabtree Neil Vyse
Buckinghamshire County Council	Nick Corby
CABE	Jon Rouse
Cornwall County Council	Kate Dixon Paul Hooper
JC Decaux	Pierre Jeanjean Martin Stephens Ajay Spolia Kim Axelsen
Dunstable Highways Agency	Geoff Chatfield
St Edmundsbury Borough Council	Ian Poole

Highways Agency	Philippa Allen
	Mima Garland
	Liz Pendall
Leeds City Council	Maurice Nixon
	Eddie McDevitt
Belle Isle Housing Office	Linda Helen
London Borough of Islington	Angela Diamandidou
London Borough of Sutton	Peter Wardill
	Steve Shew
Newcastle City Council	Paul Fenwick
Nottingham City Council	Mick Jarvis
	John Tedstone
Shropshire County Council	Robert Surl
John Simpson & Partners	Steven Walker
Stockport Metropolitan Borough Council	Mary Clarke
	Andrew Suggett
TRL Limited	Paul Forman
Unit of Urban Environment, City of Copenhagen, Denmark	Jens Ole Juul
URS Carillion	Peter Smith

Photographs of Marseille courtesy of Ville de Marseille

Photographs for Belle Isle Road, Leeds; Coldharbour Farm, Aylesbury; West Street, Dunstable: EDAW

All other photographs: Alan Baxter & Associates

Report compiled by David Taylor, David Orr, Robert Thorne and Jaimie Ferguson of Alan Baxter & Associates, assisted by Sandra Roebuck and Yann Leclercq of EDAW.